Jerome A. Berson

Chemical Discovery and the Logicians' Program

Further Titles of Interest:

J. Berson

Chemical Creativity

Ideas from the Work of Woodward, Hückel, Meerwein, and Others

ISBN 3-527-29754-5

F. Vögtle, J. F. Stoddart, M. Shibasaki (eds.)

Stimulating Concepts in Chemistry

ISBN 3-527-29978-5

C. Reinhardt (ed.)

Chemical Sciences in the 20th Century

ISBN 3-527-30271-9

C. Djerassi, R. Hoffman

Oxygen

A Play in 2 Acts

ISBN 3-527-30413-4

Jerome A. Berson

Chemical Discovery and the Logicians' Program

A Problematic Pairing

WILEY-VCH GmbH & Co. KGaA

Prof. Dr. Jerome A. Berson
Department of Chemistry
Yale University
225 Prospect Street
New Haven, CT 06520-8107
USA

This book was carefully produced. Nevertheless, authors, editors and publisher do not warrant the information contained therein to be free of errors. Readers are advised to keep in mind that statements, data, illustrations, procedural details or other items may inadvertently be inaccurate.

Cover illustration: G. Schulz, Fußgönheim (after a drawing by Etienne Carjat)

Library of Congress Card No. applied for.
British Library Cataloguing-in-Publication Data:
A catalogue record for this book is available from the British Library

Bibliographic information published by Die Deutsche Bibliothek
Die Deutsche Bibliothek lists this publication in the Deutsche Nationalbibliografie; detailed bibliographic data is available in the Internet at <http://dnb.ddb.de>.

© 2003 WILEY-VCH Verlag GmbH & Co. KGaA, Weinheim

Printed on acid-free paper

All rights reserved (including those of translation into other languages). No part of this book may be reproduced in any form – by photoprinting, microfilm, or any other means – nor transmitted or translated into a machine language without written permission from the publishers. Registered names, trademarks, etc. used in this book, even when not specifically marked as such, are not to be considered unprotected by law.

Composition: Typomedia GmbH, Ostfildern
Printing: Strauss Offsetdruck GmbH, Mörlenbach
Bookbinding: Großbuchbinderei J. Schäffer GmbH & Co KG, Grünstadt

Printed in the Federal Republic of Germany.

ISBN 3-527-30797-4

To my family

Table of Contents

Preface *XIII*

1 **Introduction** *1*
1.1 Motivations *1*
1.2 Scientific Method. Chemists and Philosophers *2*
1.2.1 Does a Scientific Method Exist? *4*
1.2.2 Epistemology of Science *5*
1.3 The Interaction of Methodological Theory and Scientific Practice *6*
1.3.1 The Case for Case Studies *6*
1.3.2 Empiricism and Naturalism *8*
1.4 Some Philosophical Schools *10*
1.5 References and Notes *12*

2 **Theories Built Up From Observations. The Inductivist Ideal** *14*
2.1 Theory: What Is It and Why Do We Need It? *14*
2.2 Theory: Where Does It Come From? *15*
2.3 Francis Bacon (1561–1626) *16*
2.3.1 Bacon and Experiment *16*
2.3.2 Bacon and Induction *17*
2.3.3 Bacon and "Crucial Experiments" *18*
2.4 Pierre Duhem and the Ideal of a "Natural Classification" *19*
2.4.1 Duhem's Criteria for a Physical Theory *19*
2.4.2 Duhem's Ideas on True and False Theories *21*
2.5 Duhem's Theory of "Natural Classification" *21*
2.5.1 Duhem, "Conviction" and "Reasons of the Heart" *22*
2.6 References and Notes *24*

3 **Theories from Anywhere: Popper's Philosophy of Conjectures and Refutations** *26*
3.1 Popper *26*
3.2 Falsificationism *27*
3.3 Where Do the Conjectures Come From? *27*
3.4 Falsificationism and the Working Scientist *28*

3.5 Popper's Intent: Descriptive Demarcation or Prescriptive Methodology? 29
3.6 Corroboration vs. Refutation in Experimental Tests of Theory. Popper's Rules 31
3.7 The Superiority of Refutations 32
3.8 Tests of Theory. Refutations and Internal Contradictions 34
3.8.1 Logical Inference and Logical Contradiction 34
3.9 Paul Feyerabend and a Stance "Against Method" 36
3.10 "Science is an Essentially Anarchic Enterprise." History Compels Ruthless Opportunism 37
3.11 Lakatos, Feyerabend, Kuhn, Hanson: Experiment is Unavoidably Entwined with Theory. The Problem of Incommensurability 38
3.11.1 Perceptions and Contexts 39
3.11.2 Prout's Theory 42
3.12 Where Do We Go From Here? 43
3.13 References and Notes 44

4 Refutation by Internal Contradiction. Kekulé's "Hypothesis of Embarrassment" and the Theory of Benzene 47
4.1 Internally Contradictory Theories Are Prohibited 47
4.2 Popper Prohibits *Ad Hoc* Rescue of Refuted Theories 47
4.3 Kekulé's Benzene 50
4.4 Objections from Baeyer, Körner, and Ladenburg 53
4.5 Tests of Structural Theory by the Number of Isomers 54
4.6 Kekulé's *Ad Hoc* Hypothesis 58
4.7 Violations of Popper's Restrictions 62
4.8 Kekulé's Theory: Success Despite Internal Inconsistency 63
4.9 Falsifiability in Principle 64
4.10 Reflections 65
4.11 Acknowledgement 66
4.12 References and Notes 66

5 Are Crucial Experiments Conclusive? Refutations: Permanent, Temporary, and Virtual 68
5.1 Refutations and "Crucial" Experiments 68
5.2 Death and Resurrection of Theories. Temporary Refutations 69
5.2.1 The Hypothesis of Non-Vicinal Hydrogen Shift in the Racemization of Camphene 69
5.2.2 A Challenge from Houben and Pfankuch 72
5.2.3 Death and Resurrection of Theories. The Non-Vicinal Shift Theory Comes Back 76
5.2.4 Further Resurrections 77
5.3 A Premature Death Certificate for Willstätter's "Small Molecule" Theory of Enzymes? 78
5.3.1 What Are Enzymes? Willstätter's Small Molecule Theory 79

5.3.2 Crystallization of Enzymes. Enzymes Are "Pure" Proteins *80*
5.3.3 An Incomplete Refutation. How Pure Are Proteins? *84*
5.4 Conclusion *88*
5.5 Acknowledgement *88*
5.6 References and Notes *89*

6 Farewell to Stasis, Welcome to Metamorphosis: Refutation of Kekulé's Rule of Minimal Structural Change in the Discovery of Molecular Rearrangements *91*
6.1 Kekulé's Rules *91*
6.2 The Pinacol Rearrangement. Fittig's "Discovery" *93*
6.2.1 Early History of the Pinacol Rearrangement *94*
6.2.2 The Atomic Weight Controversy and the Rise of Structural Theory *94*
6.2.3 The Structure of Acetone. Uncertainty Piled on Obscurity *95*
6.2.4 Fittig Attempts to Alkylate Acetone *96*
6.2.5 Rudolf Fittig (1835–1910) *97*
6.2.6 The Acetone to Pinacol Reaction *97*
6.2.7 Conflict with Städeler. A Pillow-Fight at Midnight in a Dark Cellar Between Blindfolded Combatants *98*
6.2.8 The Transformation of Pinacol to Pinacolone *99*
6.2.9 Loyalty and Dependency. Fittig's Atomic Weight Problem *99*
6.2.10 The Structures of Pinacol and Pinacolone. Friedel and Finally Butlerov *102*
6.3 The Benzilic Acid Rearrangement *108*
6.3.1 Liebig vs. Laurent: A Fortuitous Observation and a Questionable Criticism *108*
6.3.2 The Benzil-Benzilic Acid Series. Futile Attempts to Assign a Common Skeletal Structure *109*
6.3.3 The Rearrangement: Zincke Sees It But Can't Believe It. Baeyer Finally Gets It *111*
6.4 What Is a Discovery? *113*
6.4.1 Implications of the Rearrangements for Kekulé's Rules of Chemical Change *114*
6.4.2 Why Worry About Proper Attribution of Credit? *114*
6.5 Appendix. Ceteris Paribus or the Problem of Provisos: A Quiet, Menacing Relative of Theory? *116*
6.5.1 Does *Ceteris Paribus* Threaten the Completeness of Science? *116*
6.5.2 Provisos, Newton's Laws, and Planetary Astronomy *117*
6.5.3 Provisos in the Elucidation of Molecular Structure *117*
6.5.4 Can Theories Be "Complete?" *119*
6.5.5 Fruitful and Fruitless Disregard of Provisos *119*
6.6 References and Notes *120*

7 Some Non-Refutative Motivations in Science *122*
7.1. Science Without Refutation *122*
7.2 Correlation: The Search for Regularity in Nature *122*
7.3 Accidental and Serendipitous Discoveries *123*
7.3.1 Louis Pasteur and Chirality *124*
7.3.2 Wöhler and the Synthesis of Urea from Ammonium Cyanate *124*
7.4 Are Accidental Discoveries Rare? *126*
7.5 Organic Synthesis, a Corroborative Enterprise *128*
7.5.1 Synthesis of Molecules to Test Theories of Chemical Bonding and Reaction Mechanism and the Synthesis of Natural Products *128*
7.6 Reflections *131*
7.7 References and Notes *137*

8 False but Nevertheless Fruitful Theories. Speculations on the Biogenesis of Alkaloids *133*
8.1 Early Theories of Alkaloid Biogenesis *134*
8.1.1 Alkaloids from Proteins: Pictet's Theory *135*
8.1.2 The Mannich Reaction, a Key to Synthesis and Biosynthesis *136*
8.1.3 Winterstein and Trier *138*
8.2 Robinson: The Mannich Reaction, Tropinone, Biomimetic Synthesis, and Biogenesis of Alkaloids *140*
8.3 Reaction Mechanisms in the Early 20th Century *145*
8.4 From "the Mechanism of the Phytochemical Synthesis of Certain Alkaloids" to "Structural Relations of Plant Products" *146*
8.5 The Strychnine Story *147*
8.5.1 Structure *147*
8.5.2 A Digression on Structural Elucidation *148*
8.5.3 The Endgame of the Strychnine Structural Problem *150*
8.5.4 Credit and Priority *154*
8.6 Speculations on the Biogenesis of Strychnine and Other Indole Alkaloids *155*
8.6.1 The "Woodward Fission" *156*
8.7 Speculations on the Biogenesis of Other Alkaloids *157*
8.7.1 Emetine *158*
8.7.2 Cinchona Alkaloids *159*
8.7.3 Other Alkaloids *160*
8.8 The Synthesis of Strychnine *161*
8.9 The Experimentally Determined Biogenesis of Alkaloids *163*
8.9.1 Benzylisoquinolines and Pyrrolidines *163*
8.10 Tests of the Biosynthetic Role of the Woodward Fission in the In Vivo Formation of Strychnine and Other Alkaloids *165*
8.10.1 Alternatives to the Woodward Fission *166*
8.10.2 The Wenkert-Thomas Hypothesis *168*
8.10.3 Loganin and Secologanin, Key Intermediates from Mevalonate, not from Amino Acids *172*

8.11 The Startling Accuracy of a False Theory *173*
8.11.1 Popper on Replacement of a Theory *173*
8.12 Dreamers and Practitioners *174*
8.13 References and Notes *174*

9 Epilogue *178*
9.1 Where Have We Been? *178*
9.2 The "Logicians' Program" *179*
9.3 Which Way Now? *180*
9.4 References and Notes *181*

Index of Authors *183*

Index of Names *187*

Index of Subjects *191*

Preface

Many individuals have contributed to this book. Some of the ideas were discussed with colleagues and students at the seminar sessions *Issues in Science and Humanities* at Yale. Preliminary versions of some of the chapters were presented in a series of lectures I gave at the University of Colorado, Boulder, in the summer of 2002. I am grateful to Professors G. Barney Ellison, Charles DePuy, and Josef Michl for arranging that program. They and their colleagues engaged me in stimulating, instructive, and enjoyable discussions that helped greatly to clarify and strengthen my presentation of this material. The Department of Chemistry of that university provided space and facilities as needed.

Others who helped shape what you are about to read include Joseph S. Fruton, Roald Hoffmann, Frederic L. Holmes, Ruth Marcus, J. Michael McBride, Kurt Mislow, Alan J. Rocke, and Jay S. Siegel. They pointed out many potholes on my bumpy road into an unexplored region. My wife Bella again has given generously of her time, patience, and insight. The editor, Frank Weinreich, provided helpful guidance.

I thank all of these people. No doubt there may be others I have forgotten, and if that is so, it is by oversight rather than malice.

Permission to reproduce textual and graphic materials has been kindly granted by Wiley-VCH GmbH, W. H. Freeman and Company, ProQuest's Chadwyck-Healey Online database, the Edgar Fahs Smith Collection of the University of Pennsylvania, the Chemical Heritage Foundation, and the Nobel Foundation.

New Haven, Connecticut
February, 2003

1
Introduction

"It is only through science and art that civilization is of value. Some have wondered at the formula: science for its own sake; and yet it is as good as life for its own sake, if life is only misery; and even as happiness for its own sake, if we do not believe that all pleasures are of the same quality ...

Every act should have an aim. We must suffer, we must work, we must pay for our place at the game, but this is for seeing's sake; or at the very least that others may one day see."
Henri Poincaré, 1913 [1;2]

1.1
Motivations

Science as a "game" is a notion not unique to Poincaré. As we shall see, for example, in our discussion of the work of Karl Popper and his critics (Chapter 3), this metaphor is present, implicitly or explicitly, throughout the development of philosophy in the modern scientific era.

If science is a game, then, it must have rules of procedure. These rules are commonly referred to as "the scientific method," which in one definition is simply "the logic of science." [3;4] The present book examines recent formulations of these rules and compares them with events in the history of chemistry. Have chemists played by these rules? Or does progress in chemistry go on despite fundamental violations of them or even in virtual ignorance of them? That is to say, we now have the methodologists' epistemological analyses that tell us how science *has been and should be done*. To what degree does congruence exist between that body of knowledge and the facts of history, that is, how science *was done?*

To the scientist, especially to the chemist, [5;6] history and philosophy are extra-technical modes of knowledge and thought, which most of us scarcely consider in our daily work. Yet, we may be depriving ourselves of helpful insights by thus shutting them out of our consciousness. This book, like an earlier one, [7] aims to expand scientists' acquaintance with these classical branches of learning and with their power to organize events and ideas into fruitful patterns. The objective is to ascertain whether history and philosophy might directly help to increase our cognitive prowess in science itself.

At first glance, this might appear to be an attempt to smuggle humanities into science and thereby somehow to make scientists broader, more cultured, more attuned to the world at large. It is, of course, a familiar complaint that over time, the sciences have become remote from the humanities in many ways. Students in science are educated differently from those in humanities, the scholarly goals are approached by disparate methods, and communication across the gap between what C.P. Snow famously called the "two cultures" has become ever more difficult. However, I do not aspire primarily to bridge the two cultures, even though these writings may make some incidental contribution to that aim. Rather, I hope to persuade scientists, and particularly chemists, that a consideration of certain philosophical issues can make their science more insightful and more self-aware by placing it in the perspective of history.

Following two complete chapters on relevant aspects of modern philosphy of science, the main portion of the book consists of case studies of discoveries in chemistry. These have been chosen to illustrate several important philosophical issues. I hope that these will entertain and enlighten even readers who have had no previous interest in the philosophy of science.

The level of technical competence needed to follow the chemical arguments of this book is about that of a student who has had or is taking an advanced course in organic chemistry beyond the first year. Typically, junior or senior undergraduates have had such preparation. However, this is not a textbook or reference book of chemistry. Accordingly, the reader should not expect to find topical discussions of the current status of all the chemical subjects discussed here. Similarly, the coverage of subject matter is necessarily selective. Those expecting up-to-date treatment of their favorite research field or of the work of their personal scientific hero may be disappointed.

Let me also make clear that this is not a textbook of philosophy or even of the branch of it known as the scientific method. Readers should not expect a broad survey of those subjects, which have been treated in authoritative detail by professional philosophers. [8] Nor do I attempt to come to grips explicitly with several prominent problems of philosophy of science. These include the identification of truth, the problem of reduction, and the nature of reality. Recently, the last point has become entangled with the controversy over whether science reflects some aspect of the world, or instead is simply another system for gaining and exerting power over others. Discussions of some of those issues can be found elsewhere. [6;9;10] Our focus instead will be on pragmatic questions of the actual conduct of research.

1.2 Scientific Method. Chemists and Philosophers

What is the scientific method, and how do we execute its precepts? Do we follow a set of standard procedures, a check list of requirements that must be fulfilled before the research is deemed acceptable? If so, who decides what should be on the list? If not, how do we know what is a genuine contribution to our knowledge of nature?

Here is what Peter Medawar has to say about "the scientific method": [11]

> "Ask a scientist what he conceives the scientific method to be, and he will adopt an expression that is at once solemn and shifty-eyed: solemn because he feels he ought to declare an opinion; shifty-eyed because he is wondering how to conceal the fact that he has no opinion to declare."

My aim then is to give this kind of scientist some introduction to the subject from the viewpoint of one working chemist. Philosophers may object, as one already has in a review of a prospectus of this book, that given my background, this is a feckless enterprise. I am a chemist, not a historian or philosopher of science, so the issue of my qualifications (or lack of them) is obviously a concern. This point already has been raised by a reviewer [12] of my book *Chemical Creativity*, [7] a collection of studies in the history of chemistry. He correctly pointed out that I was "practising history without a license."

Although some historians may exclude that book from the realm of history, I am not so much concerned with its classification. However, a legitimate question is: Does a professional scientist have anything significant to say about the interaction of science with the disciplines of history of science or philosophy of science? How can a professional scientist, untrained – or at best marginally self-trained – in philosophy presume to debate such issues? There are two justifications.

First, I contend that in the effort to describe the fundamental activities, the thought processes, the stumbles as well as the forward steps and the reasons for both – the history as we have lived it, if you will – of a scientific field, a professional scientist has special skills to apply.

Second, the procedure I intend to use examines the *consequences* of philosophical systems. Not surprisingly, given my background, this has close analogies with something scientists do naturally, which is to test theories. At the time a theory is put forward, we may not fully understand the detailed derivation of its postulates, but if it makes clear predictions or has clearly recognizable consequences, we can compare these with experiment. Similarly, we can think of philosophical guidelines to research as analogues of theories, and we can confront them with historical events as analogues of experimental facts.

The situation is parallel to that of chemists in the early days of the application of the quantum theory to chemistry. Even if only a few chemists at that time (say approximately 1930–1960) knew much about the fundamental physics of quantum theory, its explanations and predictions relevant to our subject were sufficiently explicit to permit them to be tested by experiment. In this way, chemists were able to develop a set of attitudes toward the quantum theory, and the conclusions derived from it became incorporated into chemical usage, long before a deep theoretical training became widespread. Anyone who has tried to teach quantum mechanics to a class of students in chemistry knows that for many of them, it is easier to start with the case history of the practical chemical problem (the structure of benzene, the stereochemistry of the nucleophilic displacement reaction, etc.), rather than with a rigorous exposition of the wave equation and the derivation of matrix mechanics. Analogously, I consider my role here to be the introduction of the chemical community to some currently influential versions of scientific method by illustrating them with case histories from the chemical literature.

It might be objected – in fact, it has been argued by a reviewer of the prospectus to this book – that many of the philosophical topics I discuss here are old news to philosophers. In my opinion, this view is questionable, since many of those venerable ideas continue to be hotly debated. In any case, I have proceeded on what I believe to be a well justified assumption, namely that most chemists will not be familiar with the debates on the methodology of science that have occupied the attention of philosophers over the last five or six decades. No doubt a discussion of the up-to-the minute themes that occupy philosophers today would need a book of a very different type, but I do not consider the production of such a treatment to be part of my obligation. Instead, I propose to bring to the chemical community some of the intellectual content of those "old news" debates, especially of the ideas that have had significant impact on the way scientists perceive their own activities and discoveries.

1.2.1
Does a Scientific Method Exist?

You might be as surprised as I was to find philosophers from opposite extremes of the argument about this issue who conclude, or at least consider the possibility, that there *is no* scientific method. Karl Popper, for example, states [13] that

> "As a rule, I begin my lectures on Scientific Method by telling my students that scientific method does not exist. I add that I ought to know, having been, for a time at least, the one and only professor of this non-existent subject within the British Commonwealth."

Despite this provocative rhetorical flourish, what Popper actually claims to have done is to *demolish* the intellectual basis of earlier systems of scientific method and substitute one of his own.

Similarly, Paul Feyerabend, one of Popper's sternest critics, nevertheless does not oppose the idea that no scientific method exists. In fact, his famous (some would say notorious) book entitled *Against Method* [14] argues that all systems of scientific method, especially Popper's, are flawed and unworkable. They act as impediments to scientific progress, and therefore, he says, no scientific method *should* exist. Feyerabend summarizes his recommendation for how one should think about the problem as "anything goes." [15] Nevertheless, Feyerabend, like Popper and many other philosophers, is deeply concerned with the growth of science and with the means by which it is achieved. It will be more profitable for us to concentrate on what these philosophers *do*, rather than on what they claim *not to do*.

At the outset, my attitude was indeed shaped by the questions we have just considered: If there is a scientific method, I don't know what it is, and it seems clear that the philosophers themselves don't agree on what it is. But there is also another source of discomfort. If there is a "scientific method" but we don't understand it, can we be sure that our work is not fatally flawed? Philosophers have warned about this before. Especially during the 20th century, they have developed methodological guidelines derived from formal logic and have attempted to show that the events, discoveries, theories, controversies of science – in short, the elements of the history of science –

correspond to such guidelines. In fact, as is discussed in Chapter 9, these attempts to systematize the procedures of science have been called “the logicians’ program” for the philosophy of science. [16] One purpose of this book is to give a scientist’s perspective on these developments.

You might ask: Suppose after nine chapters worth of investigation, we conclude that the methodologists’ claims are false and do not correspond to the facts of history. Won’t we have to admit that this whole enterprise was just a waste of time? I would answer: Not necessarily.

First, the fact that philosophy may not yet have found a system that accurately describes the activities in the teeming anthill of science does not preclude the possibility that such a system may be out there waiting to be created or discovered. If we end up with a critical attitude toward existing methodology of science, perhaps some perceptive philosophers will listen to our criticisms and try again. If they are successful, we can claim to have been the catalyst for an advance in human understanding. That, if you will, is a perhaps presumptuous but at least altruistic motive for our inquiry.

Second, and closer to home, even the analyses and procedural structures the methodologists have provided so far offer us a different view of our work than we are are accustomed to. By making an effort to align the history of our discipline with the intellectual grid of a philosophical system, even a flawed one, we become more aware of elisions, assumptions, faults in reasoning, self-deceptions, and other potential ways we can blunder in our own present-day work. To use a phrase from the 1970s, we raise our level of consciousness about what we are doing.

1.2.2
Epistemology of Science

Scientific method can be considered a branch of epistemology, the study of the limits and validity of the basis of knowledge. Mathematicians and physicists (for example, Descartes, Mach, Poincaré, Weyl, Einstein, Bohr, etc.) have grappled with the problems of epistemology and, more generally, of philosophy, but few chemists [9;17,18] seem to have made such attempts. This is consistent with my own impression: most of my colleagues among working chemists seemed to be as oblivious to this activity as I was.

In this book, the problem of epistemology is approached is three stages: First, understanding some modern philosophers’ analyses of certain historical events in science; second, application of such analytical thinking to specific cases in the history of chemistry; and third, evaluation of the utility of philosophers’ formally derived guidelines in the conduct of our own scientific research. This discussion of the problematic relationship of scientists and philosophers culminates in practical questions: Should scientists be guided in their own work by normative statements of philosophers? In our research, should we follow procedures and restrictions derived from, or obedient to, propositional logic and other epistemological techniques?

1.3 The Interaction of Methodological Theory and Scientific Practice

Comparisons of philosophical guidelines with historical events in science are not new. In fact, in the contentious history of the methodology of science, many philosophers justify the validity of their logically derived proposals by reference to examples taken from scientific events of the past. However, these justifications are themselves controversial, and the advocates of rival methodologies frequently offer conflicting interpretations of the very same historical events.

These difficulties seem to have led to a major debate among philosophers over the proper relationship between the development of the theory of scientific method and the facts of scientific history. Should the "scientific method" be based upon the facts of history and their interpretation, [19] or does there exist, perhaps as an ultimate goal, a separate methodology to be elicited purely from logic and not necessarily consistent with the history of science itself? [20] Since I propose to use case studies in scientific history to test the validity of philosophical proposals, I find myself, willy-nilly, on one side of this issue. Before proceeding, I must determine whether philosophers themselves already have abandoned historical case studies as useful reference points. What follows is a brief (but I hope not distorted) outline of the main points of contention.

1.3.1 The Case for Case Studies

Hesse [19] proposes that not only should philosophers use historical cases in the development and refinement of their methodology, but also that scientists should give attention to methodological considerations in their work:

> "Firstly, a logic of science differs from a descriptive study of methodology, whether historical or contemporary, since it should supplement mere description with normative considerations. This is because it presupposes that there are norms or criteria of 'good science' in terms of which scientists judge scientific theories, and that these have some elements, perhaps tacit, of internal logical coherence and rationality ... the relation of logic and cases will be one of mutual comparison and correction."

This is an optimistic view of a problem which is pervasive in the writings in this field and is discussed further in Chapter 3: Philosophers of science do not always explicitly state whether their systems are intended as historical analyses leading to philosophical conclusions, or as normative recommendations specifying *actions* for scientists to carry out.

From the very nature of the enterprise of philosophy of science, I think the answer must be "both." In the absence of an explicit disclaimer, if the growth and success of science in the past can be analyzed according to some philosophical system, this analysis amounts to a recommendation to scientists that they adhere to the same historically proven kinds of procedure in the future. Some, but apparently not all philosophers agree with this view.

A prominent supporter, for example, is Thomas Kuhn [21] In response to Feyerabend's question [22] whether we should read Kuhn's remarks about scientific development as descriptions or prescriptions, Kuhn writes

> "The answer, of course, is that they should be read both ways at once. If I have a theory of how and why science works, it must necessarily have implications for the way in which scientists should behave if their enterprise is to flourish."

And again:

> "How could history of science fail to be a source of phenomena to which theories about knowledge may legitimately be asked to apply?" [23]

We must contend, however, with the awkward fact that the matter is not so self-evident to all philosophers. Gower, [24] for example, maintains that there is a difficulty with the concept of mutual interaction between methodology and practice, since it

> "... leaves us with no indication as to when and in what circumstances it is appropriate to treat theory (or logic) as authoritative, and when it is appropriate to treat practice (or cases) as authoritative. Scientists, faced with a task or a problem, will form a view as to what they should do in order to accomplish the task or solve the problem. Typically, this view will be shared by other scientists and, to that extent, will count as a 'received view.' But it may nevertheless be a view which conflicts with the implications of a methodological theory or logic of science. It is all very well to be told that such conflicts call for a mutual comparison and correction of logic and cases, but unless we can draw a distinction between those questions we answer by appealing to logic and those questions we answer by appealing to received opinion, the advice is useless. Short of dispensing with methodological theory altogether, the only way of resolving a conflict between theory and received opinion is by treating theory as authoritative and not allowing received opinion to modify it in any way."

This passage, in my view, reveals that at least some philosophers strongly believe there is a need for a self-referential branch of philosophy of science that is concerned primarily with the logical relationships of scientific thoughts. The "methodological theory" to which he refers seems to pertain not to the scientists' pragmatic task of how they have done or should do their work, but rather to the philosophers' concern for questions of reality, that is, for ontology.

Gower derives this austere formula from the idea that methodology based upon logic can make no compromise with experience. He admits that

> "some interaction between logic and cases may be inevitable. But the fact, if it is one, that we are obliged to temper the rigidity of a 'Euclidean' method in methodology with gestures in the direction of what we take to be real science does not mean that a relaxed 'Euclidean' method is acceptable. The way we do methodology is, perhaps, both inevitable and unacceptable."

1.3.2 Empiricism and Naturalism

In Gower's view, [25] the unresolved conflict over how a methodology of science should be developed, and what relevance, if any, the facts of scientific history can have to that enterprise, has grown out of a separation rooted in 19th century philosophy between empiricists and naturalists.

> "By empiricism is understood the view that human beings occupy a special and privileged place in our picture of the world, because it is only in terms of human experience that we can construct and understand that picture. ... For an empiricist, there is no prospect of inventing mind-independent structures and processes to explain experience, for we could not meaningfully describe them."

One finds this view poetically evoked by Poincaré: [2;26]

> "All that is not thought is pure nothingness; since we can think only thoughts and all the words we use to speak of things can express only thoughts, to say that there is something other than thought, is therefore an affirmation which can have no meaning.
>
> And yet – strange contradiction for those who believe in time – geologic history shows us that life is only a short episode between two eternities of death, and that, even in this episode, conscious thought has lasted and will last only a moment. Thought is only a gleam in the midst of a long night.
>
> But it is this gleam which is everything."

The remark "all that is not thought is pure nothingness" seems cryptic, if not actually paradoxical. In this passage, Poincaré seems to be coming perilously close to denying the existence of reality or to asserting the relativity of all knowledge. I say "paradoxical" because it is hard to believe that he really means that reality is an illusion, or that there is no "truth" beyond what we think it is or say it is, when on the contrary, his writings [2] make clear that he is a mainstream searcher after truth by the methods of science:

> "But what we call objective reality is, in the last analysis, what is common to many thinking beings, and could be common to all; this common part, we shall see, can only be the harmony expressed by mathematical laws. It is this harmony then which is the sole objective reality, the only truth we can attain; and when I add that the universal harmony of the world is the source of all beauty, it will be understood what price we should attach to the slow and difficult progress which little by little enables us to know it better."

What is the "it," the "harmony," he tells us we can come to know better if not some form or reflection of an ultimate reality?

Despite the contention over other issues, empiricism is the basic mind-set of much of 20th century philosophy. Nevertheless, the theories of scientifc method that have come out of the empiricist school

"... bear a relation to scientific evidence which would be regarded as intolerable in the case of scientific theories. What matters for these theories of scientific method is not the superficial record of what scientists have done and do, but the real, objective, rational standards which they attempt to identify, and which lie behind or beyond that record ... We do not judge empiricist theories of scientific method – logics of science – by empirical evidence; rather we base their warrant upon decisions concerning the nature of the rationality which, we think, must lie behind and must explain the remarkable achievements of science." [27]

I interpret this passage to mean that there is a logic of scientific method which can be derived independently of the facts of scientific history; even if philosophers cannot agree on what the "standards" are, some of them think such standards must exist or must be discoverable.

My inspection of the philosophical writings of such scholars as Popper, Kuhn, Lakatos, and Feyerabend, however, leads me to think that, on the contrary, many 20th-century leaders of philosophy of science have not hesitated to use examples from history to justify their theories of scientific methodology, even though theorists of rival, mutually incompatible methodologies frequently treat the same set of historical facts by diametrically opposing interpretations. Lakatos, [28] for example, writes that it is simply *hubris* to demand that if, say, Newtonian or Einsteinian science turns out to have violated Bacon's, or Carnap's, or Popper's *a priori* rules of the game, the business of science should be started anew.

Naturalism, [27;29] in contrast to empiricism, asserts that

"... human beings, together with their capacities and abilities, belong to an objective natural order, and ... we should explain their relation to the world in terms appropriate to that order ... There is nothing special or central about human experience which entitles us to treat it as authoritative." [24]

According to Gower, naturalism maintains that human observation is not privileged; hence it cannot serve to validate methodology, because

"Initial impressions or appearances can provide no more than a record of the passing show in which we discern useful generalities. Naturalism urges us to erect theories [of scientific method] which will enable us to understand such a record, to complete the gaps in it, and sometimes to correct it." [27]

Some readers may ask, as I did, since logic itself is a human construct, does not a claim of its superiority as a determinant of methodology conflict with naturalists' denial of a privileged place for human cerebration?

On the other hand, Kitcher [30] sees a clear justification for case studies:

"Ultimately, philosophical pictures should be tested against a range of cases, historical and contemporary. Articulating the test procedure is itself a formidable task – and moreover, one that cannot be detached from our philosophical picture of science."

Apparently, some philosophers think that the traditional accounts of scientific methodologies rest on an empiricist epistemology, but the logic is (or should be) justified by naturalist means. It has been suggested [29] that the achievement of a naturalist justification of a normative methodology will have to incorporate human factors. One would need to resolve such issues as how the human cognitive apparatus processes observations and arguments, and what effects human interactions may have on the growth of knowledge. This will require naturalist philosophers to incorporate psychology, sociology, and even economics into their theories of methodology, actions that would represent a radical break with the views of many empiricist philosophers.

Obviously, an amateur attempting to pursue this conflict to its resolution risks being considered either presumptuous or foolish, categorizations I hope to dodge. Nevertheless, it is my impression that the modern naturalistic program for the epistemology of science is just beginning to emerge. [29;31] The bulk of scientific methodology produced in the 20th century has an empiricist foundation. Even in its internal battles, the contestants use case studies as the ultimate test of methodology. Thus, I think I may do likewise, even while recognizing that this procedure may one day go out of fashion. Some further justification for this decision is given in Chapter 3.

1.4 Some Philosophical Schools

It should be clear by now that there is no one, universal scientific method. At the risk of oversimplification, we may think of philosophers as belonging to one or another of several contending factions. Each of these gives rise to an associated set of conclusions about the methodology of science, and each is largely incompatible with the others.

That many systems of philosophy exist, each passionately attacked by its adversaries and fiercely defended by its adherents, should come as no surprise. This state of affairs has existed at least since the time of the ancient Greeks, and disputation has been an essential element of philosophy ever since. [3;8] Therefore, it does not seem wise for us to expect to receive a finished, generally accepted set of rules for procedure any time soon. Our decision on whether to adopt one philosophical system or another, or to use elements from more than one, or to reject all methodological strictures entirely, depends upon our own resolve: we need to learn enough about philosophy to make a reasoned choice.

Another objective lies beyond that of stimulating chemists and other scientists to be conscious of these ideas. It is the hope that philosophers will be motivated (but not too irritated!) by the results of the process of matching philosophical theory to historical data to make greater use of such events from the literature of chemistry for philosophical analyses.

It will be useful to focus on two principal streams of thought about how science should be done. One is the *inductive method,* in which a major theme may be described (with some oversimplification) as the construction of theories from observations. The other is the *deductive method,* which concerns the derivation and test of predictions from theories. These two methods have generated many philosophical approaches to the rationalization and interpretation of events in the history of science. I shall not attempt to retrace that development, nor shall I explore in any detail (let alone become a participant in) the vast web of controversy that grew around them during the 20th century. Instead, I have selected aspects of that material which I have found to strike special chords of relevance to events in science I have lived through or studied. Readers who would like to pursue the philosophical basis of the recent controversies at greater depth and breadth would find authoritative guidance in other books and reviews, among which I have found especially helpful those of Lakatos, [28] Feyerabend, [14] and Gower. [8]

Some of the inferences I have drawn from my explorations of historical instances in a philosophical context are not new. Philosophers and historians of science have pointed out similar conclusions from their own researches. My justification is not novelty but persuasion: In the philosophical literature, the most frequently cited historical cases concern events in physics, astronomy, and biology. Chemists may be able to evaluate cases from chemistry more easily, and in succeeding chapters of this book, I will provide such examples. I expect that the reader of this book will find that these histories of familiar chemical concepts will be more easily understood.

Chapter 2 gives a sampling of inductive thought, as exemplified in the writings of philosophers active during the period of the development of modern science. These include Francis Bacon, Henri Poincaré, and Pierre Duhem. Chapter 3 introduces philosophers such as Karl Popper and Imre Lakatos, whose methodology has a deductivist basis and may be summarized in Popper's term "conjectures and refutations."

We examine not only the ideas of each of these groups but also the counter-arguments put forward against them by adherents of the other group and by less easily classified critics such as Kuhn, Feyerabend, and others.

Among the topics to be explored are two of the various sources that lead to the *creation* of theories: inductive derivation of theory from observation (Chapter 2) and theory from hunch, conjecture, or unspecified origin (Chapter 3). We also examine *tests* of theory by verification or by two kinds of refutation, namely by discovery of internal contradiction or by experimental falsification (Chapters 3, 4, and 5). We observe some of the factors that lead to the death and occasional rebirth of theories (Chapter 5). We study discoveries sparked accidentally, without benefit of theory (Chapter 7), and we find that some astonishingly fruitful theories can nevertheless originate in entirely false premises (Chapter 8).

In sum, the dominant question we address is this: Does methodology of science guide chemical discovery? I hope that many readers will find this exploration stimulating, provocative, and even useful. If the project ultimately is considered an exercise in popularization, so be it.

Those who remain skeptical I hope at least to bring into agreement with Byron's hero Manfred, whose own reservations about philosophy were tempered with admiration: [32]

"There is a calm upon me –
Inexplicable stillness! which till now
Did not belong to what I knew of life.
If that I did not know Philosophy
To be of all our vanities the motliest,
The merest word that ever fooled the ear
From out the schoolman's jargon, I should deem
The golden secret, the sought "Kalon," [33] found,
And seated in my soul. It will not last,
But it is well to have known it, though but once:
It hath enlarged my thoughts with a new sense,
And I within my tablets would note down
That there is such a feeling."

1.5 References and Notes

1 H. Poincaré, *Science and Hypothesis, in The Foundations of Science* The Science Press, Lancaster, PA **1913**. Reprinted 1946, G.B. Halsted, translator.
2 H. Poincaré, **1913**: p. 355.
3 B. Gower, *Scientific Method: An Historical and Philosophical Introduction,* Routledge, London, **1997**.
4 B. Gower, **1997**: p. vi.
5 This reluctance of chemists to engage philosophical issues (except when defending chemistry from its detractors) has been noted elsewhere (see the following reference, p. 1).
6 N. Psarros, *Die Chemie und Ihre Methoden: Eine philosophische Betrachtung,* Wiley-VCH, Weinheim, Germany, **1999**.
7 J. A. Berson, *Chemical Creativity: Ideas from the Work of Woodward, Hückel, Meerwein, and Others,*Wiley-VCH, Weinheim, **1999**.
8 B. Gower, **1997**: This book gives an instructive introduction to the history of scientific method.
9 N. Bhushan, S. Rosenfeld, eds., *Of Minds and Molecules,* Oxford University Press, New York, NY, **2000**. See also numerous references cited therein.
10 A. Sokal, J. Bricmont, *Fashionable Nonsense: Postmodern Intellectuals' Abuse of Science,* Picador USA, New York, **1998**. See also numerous references cited therein.
11 P. Medawar, *Induction and Intuition in Scientific Thought* Methuen, London **1969**. 11.
12 J. Labinger, *Science* **1999**,*285,* 2075.
13 K. R. Popper, *Realism and the Aim of Science: From the Postscript to the Logic of Scientific Discovery* (Ed.: W. W. Bartley), Rowman and Littlefield, Totowa, NJ **1983**. p. 5.
14 P. Feyerabend, *Against Method: An Outline of an Anarchistic Theory of Knowledge,*New Left Books, New York, **1975**. Reprinted as *Against Method* in 1997 by Verso from the third edition of 1993.
15 P. Feyerabend, **1975**. p. 14ff.
16 S. Toulmin, *From Logical Systems to Conceptual Populations PSA 1970 In Memory of Rudolf Carnap* (Eds.: R. C. Buck, R. S. Cohen), D. Reidel, Dordrecht, The Netherlands **1971**. pp. 552ff.
17 M. Polanyi, *Personal Knowledge; Towards a Post-Critical Philosophy,* Routledge & Kegan Paul, London, UK, **1958**.
18 E. F. Caldin, *The Power and Limits of Science; a Philosophical Study,* Chapman & Hall, London, **1949**.
19 M. Hesse, *The Structure of Scientific Inference* University of California Press, Berkeley, CA **1974**. p. 6.
20 B. Gower, **1997**: p. 257.
21 T. S. Kuhn, in ref. 22, p. 237.
22 P. Feyerabend, in *Criticism and the Growth of Science* (Eds.: I. Lakatos, A. Musgrave),

Cambridge University Press, Cambridge, U.K. **1999**. p. 198.

23 T. S. Kuhn, *The Structure of Scientific Revolutions*, University of Chicago Press, **1970** ed. p. 9.

24 B. Gower, **1997**: p. 258.

25 B. Gower, **1997**: pp. 236ff.

26 Poincaré, I think, is not to be interpreted in this passage as denying the existence of reality or of asserting the relativity of all knowledge. He does not believe that reality is an illusion, or that there is no "truth" beyond what we think it is or say it is. On the contrary, his writings make clear that he is a mainstream searcher after truth by the methods of science.

27 B. Gower, **1997**: p. 259.

28 I. Lakatos, in *PSA 1970 in Memory of Rudolf Carnap* (Eds.: R. C. Buck, R. S. Cohen), D. Reidel, Dordrecht, Holland **1971**. p. 121. Boston Studies in the Philosophy of Science, Vol. VIII.

29 P. Kitcher, *Phil. Rev.* **1992**, *101*, 53.

30 P. Kitcher, **1993**: pp. 9–10, fn 14.

31 P. Kitcher, *The Advancement of Science:Science without Legend, Objectivity without Illusions*, Oxford University Press, Oxford, **1993**.

32 Baron George Gordon Byron, *Manfred*, in *The Works of Lord Byron* (E. H. Coleridge [and R. E. Prothero], eds.), new, revised, and enlarged edition, John Murray, Charles Scribner's Sons, London, UK, 1898–1904. From ProQuest's Chadwyck-Healey Literature Online database. Reproduced with permission.

33 *Webster's New International Dictionary*, 2nd ed. G. & C. Merriam, Springfield, MA, **1959**. Kalon: a noun, from the Greek, meaning "the beautiful, – frequently with an implication of moral as well as aesthetic beauty."

2
Theories Built Up From Observations. The Inductivist Ideal

"The scientist must set in order. Science is built up with facts, as a house is with stones. But a collection of facts is no more a science than a heap of stones is a house."
J.H. Poincaré, 1913 [1;2]

2.1
Theory: What Is It and Why Do We Need It?

Working scientists comprehend and systematize their knowledge of the world in mental constructs called theories or hypotheses. (I make no distinction here between the two). Theories come and go, and we might ask, with Josiah Royce: [3]

> "Why, then, does science actually need general theories, despite the fact that these theories inevitably alter and pass away? What is the service of a philosophy, when it is certain that the philosophy of science which is best suited to the needs of one generation must be superseded by the advancing insight of the next generation? Why must that which endlessly grows, namely, man's knowledge of the phenomenal order of nature, be constantly united in men's minds with that which is certain to decay, namely the theoretical formulation of special knowledge in more or less completely unified systems of doctrine?"

One answer comes from a recognition of the several levels of our knowledge of natural phenomena. Roughly, these may be identified with a set of raw observations, then a law, which expresses and codifies related observations into identifiable groups, and finally a theory that groups together or classifies related laws, explains why they are related, and predicts new phenomena. For the present then, we can follow Mach [4] as quoted by Duhem [5;6] and think of one of the purposes of theory as the achievement of an "economy of thought."

Some theories are widely accepted and guide the thinking and action of most of the scientific community, some are accepted as probable, although acceptance may not be universal, some are speculative, and some, although currently accepted, are destined to be shown wrong.

Thus, the scientist recognizes that theories are not only predictive but also pragmatic and vulnerable. In Poincaré's vivid formulation: [1;2]

> "Thanks to generalization, each fact observed enables us to foresee a great many others; only we must not forget that the first alone is certain, that all others are merely probable. No matter how solidly founded a prediction may appear to us, we are never *absolutely* sure that experiment will not contradict it. The probability, however, is often so great that we may be content with it. It is far better to foresee even without certainty than not to foresee at all."

Beyond these methodological significances, there lies a more ethereal realm in which theory "connotes not merely an abstract, systematic understanding of the world, but one in secret communion with the gods," an idea that can be traced back to Plato. [7] [8] Indeed, the very word *"theory,"* its Greek antecedents *theōria* (meaning a spectacle or speculation), and *theōrein* (to look at), all are derived from the same root as the Greek *theos,* (god).

The creation and testing of theories lie at the core of science and define its program. In this sense, the experimentalist treasures theory as the source of inspiration for new investigations.

What theories are and how they are created is a subject so broad that its boundaries may be considered coterminous with those of the philosophy of science itself. For pedagogical purposes, we limit our discussion to the views of a few key figures in the history of philosophy. My choice does not imply any assignment of relative importance or influence. Rather, I intend to illustrate how some of the leading philosophical ideas can serve as guides in the analysis of historical events in chemistry.

2.2 Theory: Where Does It Come From?

We may distinguish two major currents in the creation of theories, or at least, in the claims their authors make for their origins. One, which we may loosely call the inductivist stream in science, flows out of many experimental observations and represents the generalization and explanation of experience.

The second stream, which we may call *falsificationism,* or the method of *conjectures and refutations,* is a fundamentally deductivist methodology. It denies the legitimacy of the basic premises of inductivism and pays scant attention to the origins of theories. Instead, it argues that science moves forward by critical tests designed to *refute* theories, whatever their origin.

In this chapter, we trace some of the origins and guiding principles of inductivism. [9] Later chapters will deal with falsificationism and some of its implications.

2.3
Francis Bacon (1561–1626)

Bacon is often credited with the formulation of the first precepts of what came to be known as the *inductive method* in science. It is worth noting that, although he was a man of many talents – writer, lawyer, statesman, and adviser to Elizabeth I and to James I of England – he was not a professional scientist. Indeed, in Bacon's day, the community of scholars devoting themselves to the understanding of the natural world was miniscule. How this man came to exert such a strong influence in the early modern period of science and beyond makes a fascinating story, which others have told. [[10–13]

2.3.1
Bacon and Experiment

In Gower's words

> "The importance of Bacon's contribution lies ... in the encouragement he gave to the experimental style in science. Real practical experiments were, in his view, indispensable to an understanding of natural processes." [14;15]

How did a practicing lawyer come to the advocacy of experimentation in the pursuit of theoretical insight about nature? Gower draws an interesting analogy between Bacon's placement of experimentation in science and his recognition that lawyers need cross-examinations and trials to make justice prevail, that is "the value of experiments which exhibit nature under investigation lies in their ability to reveal the truths nature would otherwise conceal." [16]

In the *Novum Organum,* written in 1620 in the very dawn of experimental science, Bacon wrote

> "Those who have handled sciences have been either men of experiment or men of dogmas. The men of experiment are like the ant, they only collect and use; the reasoners resemble spiders, who make cobwebs out of their own substance. But the bee takes a middle course: it gathers its material from the flowers of the garden and of the field, but transforms and digests it by a power of its own. Not unlike this is the true business of philosophy; for it neither relies solely or chiefly on the powers of the mind, nor does it take the matter which it gathers from natural history and mechanical experiments and lay it up in the memory whole, as it finds it, but lays it up in the understanding altered and digested. Therefore from a closer and purer league between these two faculties, the experimental and the rational (such as has never yet been made), much may be hoped."

We can discern here the kernel of the inductivist idea: Experiments and observations alone do not constitute wisdom. The mind must "alter and digest" them and ultimately understand them by means of a theory.

2.3.2 Bacon and Induction

Bacon [17] made the point that his new scientific method differed from traditional logic, whose primary activity he saw as "demonstration by syllogism." The latter takes the familiar form of a major premise (all fish swim), a minor premise (this creature is a fish), and a conclusion (therefore this creature swims). The conclusion reached by this deductive path is implied in the premises themselves, regardless of whether the premises are in fact true. Bacon characterizes this kind of reasoning as suited to disputation but not to an understanding of the natural world.

A related issue is the distinction between mathematics and empirical science. Hempel [18;19] makes clear that empirical knowledge and mathematical truth are distinct concepts, since the latter is found by

> "... the method of mathematical demonstration, which consists in the logical deduction of the proposition to be proved from other propositions, previously established. Clearly, this procedure would involve infinite regress unless some propositions were accepted without proof; such propositions are indeed found in every mathematical discipline which is rigorously developed; they are the *axioms* or *postulates* ... of the theory ... all that pure mathematics is interested in, and all that it can establish, is the deductive consequences of given sets of postulates and thus the necessary truth of the ensuing theorems relatively to the postulates under consideration."

Granted that this incongruency between the kinds of "knowing" in empirical science and in mathematics derives from the difference in axiomatic structure of the two disciplines. Nevertheless, it cannot be denied that there are strong similarities in the way the two (largely separate) communities perceive forward progress by confirmations. Thus, it has been stated [20] in reference to Hilbert's famous mathematical problem set of 1900 that

> "Hilbert's talk [to the International Congress of Mathematicians] highlighted the importance of hard, unsolved problems in the development of mathematics. Problems reveal the gaps in our knowledge, and in order to solve them mathematicians often have to construct deep and powerful abstract theories. When those theories lead to the solution of the problem that was their initial motivation, this success serves as a confirmation that we are on the right track, that we have indeed discovered something new."

So despite the difference in the epistemological origin of their aims, scientists and mathematicians both seem to need similar *psychological* reassurances.

Hempel quotes Einstein:

> "As far as the laws of mathematics refer to reality, they are not certain; and as far as they are certain, they do not refer to reality."

How different this is from the program of empirical science, which traditionally has sought to find, or at least approach, the real nature of things and phenomena! In an embryonic way, Bacon's early rejection of logic as a path to empirical truth was a response to this fundamental dichotomy.

Bacon characterizes ancient science as developing its first principles from a few examples. Taking these principles as fixed, the Greek philosophers deduced from them consequences and corollaries and ultimately predictions. [21] If new facts were then discovered to be in conflict with the predictions, the ancients explained these by subtle changes in the theory or by dismissing the observations as exceptions.

As a better path to the truths of nature, Bacon proposes induction. [22] He plans to "proceed regularly and gradually from one axiom to another, so that the most general are not reached till the last." Theories, in this procedure, will be built up from experiments, so that each higher-order conclusion will rest firmly on an unassailable base. Then, when predictions are deduced from the finally reached highest-order generalization, deviations should not be expected.

2.3.3 Bacon and "Crucial Experiments"

An idea of Bacon's which has drawn much attention describes the so-called "crucial experiment" or "instance of the fingerpost." [23] Bacon borrowed the term from

> "the fingerposts which are set up where roads part, to indicate the several directions. These I also call *Decisive and Judicial,* and in some cases, *Oracular and Commanding Instances.* I explain them thus ... instances of the fingerpost show the union of one of the natures [assumed property of a thing] with the nature in question to be sure and indissoluble, of the other to be varied and separable; and thus the question is decided, and the former nature is admitted as the cause, while the latter is dismissed and rejected. Such instances afford very great light and are of high authority, the course of interpretation sometimes ending in them and being completed. Sometimes these instances of the fingerpost meet us accidentally among those already noticed, but for the most part they are new, and are expressly and designedly sought for and applied, and discovered only by earnest and active diligence."

Similarly, he wrote of so-called "Experiments of Light" (*Experimenta Lucifera): [24]*

> "Now experiments of this kind have one admirable property and condition: they never miss or fail. For since they are applied, not for the purpose of producing any particular effect, but only of discovering the natural cause of some effect, they answer the end equally well whichever way they turn out; for they settle the question."

These formulations pointed the way toward a probing style of science, in which a decision between (or among) two (or more) explanations or theories of a phenomenon would be achieved by a carefully planned test, the predicted outcome of which would differ from one theory to another.

We see the same style at work in Bacon's criticism of Aristotle, who proposed to verify theories by the collection of affirmative instances only:

Bacon writes

> "To conclude *upon an enumeration of particulars without instances contradictory* is no conclusion, but a conjecture; for who can assure ... that there are not others on the contrary side which appear not."

This objection appears again as the basis for Popper's later assertion (see Chapter 3) that *falsification,* not verification, is the most reliable way for science to move forward.

2.4 Pierre Duhem and the Ideal of a "Natural Classification"

Almost three hundred years later, we can discern an inductivist element in the work of the French physicist and philosopher Pierre Duhem (1861–1916). [5;25] It will be useful to examine his idea that the course of development of a good physical theory tends toward the achievement of a "natural classification." So we must know his definitions of a theory and a "natural classification," and we must understand the line of reasoning that led him to propose the nature of a theory's ultimate objective. His argument, elevated in rhetoric, rich in example, and provocative in impact, cannot be appreciated fully except by study of his own presentation. [5] I venture to paraphrase it here, although occasionally I have not been able to resist the urge to quote from it.

Duhem points out that if theory were to be defined in a metaphysical or ontological way, as for example "a hypothetical explanation of material reality," one would have to define what is meant by "reality." Since there are so many conflicting philosophical views of this definition, Duhem asks:

"Could we not assign an aim to physical theory that would render it *autonomous*? Based on principles which do not arise from any metaphysical doctrine, physical theory might be judged in its own terms without including the opinions of physicists who depend on the philosophical schools to which they may belong." [26]

> "... A physical theory is not an explanation. It is a system of mathematical propositions, deduced from a small number of principles, which aim to represent as simply, as completely, and as exactly as possible a set of experimental laws." [26]

2.4.1 Duhem's Criteria for a Physical Theory

Duhem then outlines "four fundamental operations in a physical theory: (1) the definition and measurement of physical magnitudes; (2) the selection of hypotheses; (3) the mathematical development of the theory; (4) the comparison of the theory with experiment."

Despite the title of his book, *The Aim and Structure of Physical Theory,* it seems clear that Duhem intended his philosophy to apply to chemistry and other sciences as well. If that were not the case, how could one reconcile physics and chemistry? They are neighboring sciences, which blend into one another at the edges, so that it is often hard to distinguish to which discipline a particular set of observations or theoretical constructs should be assigned. It therefore would be a poor epistemology indeed that applied to one but not to the other.

Duhem was not alone among philosophers in his conviction that mathematical theory was a necessary component of theory. Henri Poincaré, the French matheticical physicist, writes [1;27]

> "But what we call objective reality is, in the last analysis, what is common to many thinking beings, and could be common to all; this common part, we shall see, can only be the harmony expressed by mathematical laws. It is this harmony then which is the sole objective reality, the only truth we can attain; and when I add that the universal harmony of the world is the source of all beauty, it will be understood what price we should attach to the slow and difficult progress which little by little enables us to know it better."

If Duhem's intention that "physical theory" should apply to chemistry is admitted, I see a difficulty in his definition of a theory as "a system of mathematical propositions," because many of the major theories of chemistry were not formulated in mathematical terms.

For example, the most prominent precursor of the structural theory, the mid-19th century "type theory" [28] was qualitative and was put forward without mathematical undergirding. It proposed the idea that chemical substances of similar "type" behaved similarly, essentially the characteristic Duhem [29] had in mind in his discussion of the classification of compounds by analogy.

The hypothesis of a guiding analogical principle in the relationship of constitution and reaction propensity in chemistry of course ultimately received a more specific basis in the structural theory. [30] The similarities of behavior were traced to similarities in *structure,* that is, in the connectivity of the constituent atoms of the molecule. Correspondingly, the concept of functional groups, reactive structural entities common to a whole set of compounds, became a major classificatory element in organic chemistry.

The theory of the classification in terms of types or structural formulas has guided the field for over a century. Its founders imparted no significant mathematical component to it. Although much later attention was given to the use of mathematical approaches to enumerate [31–33] and predict acceptable structural connectivity patterns, these did not come on the scene until long after the structural theory was well established and in use universally. In any case, such mathematical methods depended upon the structural theory itself to define acceptability.

Similarly, think of the theory of stereochemistry that developed from the van't Hoff-LeBel hypothesis of the tetrahedral carbon atom. We may ask what were the "physical magnitudes" contained in it that in Duhem's first "fundamental operation" had to be defined and measured? And what "mathematical development" (third "fundamental operation") had to take place in this theory before it became widely accepted?

Of course, it might be argued that a qualitative theory is no theory at all. This would make Duhem's requirements self-fulfilling, since, by definition, only those systems of thought that have a mathematical component could be called theories. But what is the essential distinction in function between a mathematical theory and a qualitative one? Both explain known phenomena, and both serve as stimuli for experiment. Is the amount of mathematical content to be taken as a measure of the quality of a theory? That is, the more mathematics, the better the theory? Unless one is prepared to banish great blocks of activity from the domain of science, I dare to say, although I am no philosopher or logician, that this view is not defensible.

At this point, it seems to me that there are only two alternatives: either "physical theory" does not apply to chemistry, or not all of Duhem's "fundamental operations" are required in order for a working theory to be developed and accepted. I conclude that only the second of these possibilities is valid, that is, that of Duhem's original four operations, only two are left: operation (2), the selection of hypotheses, and operation (4), the comparison of the theory with experiment. As a chemist, I now cannot quarrel with these.

2.4.2 Duhem's Ideas on True and False Theories

According to Duhem, [6]

> "... a true theory is not a theory which gives an explanation of physical appearances in conformity with reality; it is a theory which represents in a satisfactory mannner a group of experimental laws. A false theory is not an attempt at an explanation based on assumptions contrary to reality; it is a group of propositions which do not agree with the experimental laws. *Agreement with experiment is the sole criterion of truth for a physical theory.*" [6]

Here, I fear that Duhem's rhetoric has led him to overstate the case. [25] In fact, we shall see examples in which a *false* theory satisfactorily explains and even correctly predicts the outcome of certain experiments, and no doubt, the reader will be able to cite further cases. Thus, Duhem's formulation at least would have to be changed to some such form as *agreement with experiment is a necessary but insufficient criterion of truth for a physical theory.*

2.5 Duhem's Theory of "Natural Classification"

A major concept in Duhem's writings is the "natural classification", which he defines as "the ideal form toward which theory tends..." Duhem imagines the mental struggle to find the reality behind observation in the form of a metaphor: [34]

> "A slave to positive method, the physicist is like the prisoner of the cave: the knowledge at his disposal allows him to see nothing except a series of shadows in profile on the wall facing him; but he surmises that this theory of silhouettes whose outlines are shadowy is only the image of a series of solid figures, and he asserts the existence of these invisible figures beyond the wall he cannot scale."

Thus, the physicist believes in the reality of a truly ontological structure to the world, but he is forced by circumstance to approach this reality, mimicked by the "natural classification," only in stages, by successive refinements of theory. Duhem seems to be proposing that the ideal theory is a kind of asymptotic limit toward which we can aspire as the goal of our diligence. As we shall see in Chapter 3, Popper takes a much more pessimistic view of "refinements" to theory.

In a memorable analogy, [35] Duhem compares scientific progress to a mounting tide:

> "Whoever casts a brief glance at the waves striking a beach does not see the tide mount; he sees a wave rise, run, uncurl itself, and cover a narrow strip of sand, then withdraw by leaving dry the terrain which it had seemed to conquer; a new wave follows, sometimes going a little farther than the preceding one, but also sometimes not even reaching the sea shell made wet by the former wave. But under this superficial to-and-fro motion, another movement is produced, deeper, slower, imperceptible to the casual observer; it is a progressive movement continuing steadily in the same direction and by virtue of it the sea constantly rises. The going and coming of the waves is the faithful image of those attempts at explanation which arise only to be crumbled, which advance only to retreat; underneath there continues the slow and constant progress whose flow conquers new lands, and guarantees to physical doctrines the continuity of a tradition."

2.5.1
Duhem, "Conviction" and "Reasons of the Heart"

How is the achievement of this level of correspondence between theory and the underlying essence of nature to be recognized? Duhem tells us:

> "The neat way in which each experimental law finds its place in the classification created by the physicist and the brilliant clarity imparted to this group of laws so perfectly ordered persuade us in an overwhelming manner that such a classification is not purely artificial, that such an order does not result from a purely arbitrary grouping imposed on laws by an ingenious organizer. Without being able to explain our conviction, but also without being able to get rid of it, we see in the exact ordering of this system the mark by which a natural classification is recognized. Without claiming to explain the reality hiding under the phenomena whose laws we group, we feel that the groupings established by our theory correspond to real affinities among the things themselves ...
>
> The physicist cannot compel himself to believe that a system capable of ordering so simply and so easily a vast number of laws, so disparate at first encounter, should be a purely artificial system. Yielding to an intuition which Pascal would have recognized as one of those reasons of the heart 'that reason does not know,' he asserts his faith in a real order reflected in his theories more clearly and more faithfully as time goes on ...
>
> And so, we may here apply that profound thought of Pascal: 'We have an im-

potence to prove, which cannot be conquered by any dogmatism; we have an idea of truth which cannot be conquered by any Pyrrhonian skepticism.'" [36]

Reading this impassioned rhetoric, one is struck by Duhem's ultimate reliance on such psychological, even emotional, concepts as "conviction," "faith," and "reasons of the heart." But as de Broglie points out, [25] Duhem's "convictions" that a theory had reached the level of a "natural classification" sometimes were achieved only by ignoring other aspects of the phenomena explained by the theory. A case in point is given in Duhem's use of the wave theory of light as an example of a "natural classification": [36]

"But when, after much groping, we succeed in formulating with the aid of this vibration a body of fundamental hypotheses, when we see in the plan drawn by these hypotheses a vast domain of optics, hitherto encumbered by so many details in so confused a way, become ordered and organized, it is impossible for us to believe that this order and this organization are not the reflected image of a real order and organization; that the phenomena which are brought together by the theory, *e.g.*, interference bands and colorations of thin layers, are not in truth slightly different manifestations of the same property of light; and that phenomena separated by the theory, *e.g.*, the spectra of diffraction and of dispersion, do not have good reasons for being in fact essentially different."

These words were written in 1905 for the first edition of Duhem's book, *The Aim and Structure of Physical Theory*, published in 1906. The year 1905 was also the year in which Einstein showed that not all of the properties of light can be understood by the wave theory. Instead, one needs a new theory, the quantum theory, to explain the corpuscular behavior manifested in the photoelectric effect. The first edition of Duhem's book was based upon a series of articles published in the *Revue de Philosophie* in 1904–1905, Duhem then might not have been aware of Einstein's results in time to comment on them in the first edition, but by 1914, when the second edition was published, he surely could not avoid dealing with the issue. Here is his response, in the author's preface to the second edition:

"... Since that time [*i.e.*, 1906] a number of controversies concerning physical theory have been raging among philosophers, and a number of new theories have been proposed by physicists. Neither these discussions nor these discoveries have revealed to us any reasons for casting doubt on the principles we had stated. Indeed, we are rather more confident than ever that these principles should be firmly held. It is true that certain schools have affected scorn for them; free from the constraint these schools might have felt on account of these principles, they think they can run all the more easily and quickly from one discovery to another; but this frantic and hectic race in pursuit of a novel idea has upset the whole domain of physcial theories, and has turned it into a real chaos where logic loses its way and common sense runs away frightened."

Of course, Duhem is not the first (or the last!) scholar to cling to "conviction" in the face of a challenge. But our concern here is not to determine who was "right" and who was "wrong" about the nature of light. That issue has been subsumed in the paradox of the wave-particle duality, whose mysteries even now continue to challenge theorists and philosophers. The lesson that this episode has for us is this: the thought that there is an ideal "natural classification," that the rising tide of scientific progress will "conquer new lands," so ardently advocated by Duhem, itself may be an unattainable ideal. This basically pessimistic thought may not necessarily imply that there is no ultimate truth, but only that even when we reach such a stage, we may not be able to recognize it. In other words, it may be in the nature of things that all theories must be considered provisional and that we can never succeed in the wistful search for some kind of final theory. Popper [37]has expressed this view in another analogy:

> "The empirical basis of objective science has thus nothing 'absolute' about it. Science does not rest upon solid bedrock. The bold structure of its theories rises, as it were, above a swamp. It is like a building erected on piles. The piles are driven down from above into the swamp, but not down to any natural or 'given' base; and if we stop driving the piles deeper, it is not because we have reached firm ground. We simply stop when we are satisfied that the piles are firm enough to carry the structure, at least for the time being".

Our discussion of Popper's philosophical system in Chapter 3 will examine the basis of this conviction.

2.6 References and Notes

1 J. H. Poincaré, *Science and Hypothesis,* 1946 ed., G. B. Halsted, translator, The Science Press, Lancaster, PA, **1913**.
2 J. H. Poincaré, **1913**: p. 127.
3 J. Royce, **1913**: p.14–15. Introduction to Poincaré **1913**.
4 E. Mach, *The Economical Nature of Physical Research, Mach's Popular Scientific Lectures,* T. J. McCormack, Open Court, La Salle, IL, **1907**. Chapter XIII of the 3rd edition translated from the German original.
5 P. Duhem, *The Aim and Structure of Physical Theory,* P.P. Wiener, translator, Princeton University Press, Princeton, NJ, **1954**. Reprinted 1991 from the 2nd edition, Marcel Rivière, Paris, 1914.
6 P. Duhem, **1954**: p. 21.
7 S. Fuller, *Thomas Kuhn: A Philosophical History for Our Times,* University of Chicago Press, Chicago, **2000**.
8 S. Fuller, **2000**: p. 63.
9 R. Carnap, in *Philosophical Foundations of Physics,* (Ed.: M. Gardner) Basic Books, New York, **1966.**
10 B. Gower, *Scientific Method: An Historical and Philosophical Introduction,* Routledge, London, **1997**.
11 B. Gower, **1997**: pp. 40–61.
12 L. Jardine, *Hostage to Fortune: The Troubled Life of Francis Bacon,* Victor Gollancz, London, **1998**.
13 J. R. Solomon, *Objectivity in the Making : Francis Bacon and the Politics of Inquiry,* Johns Hopkins University Press, Baltimore, **1998.**
14 B. Gower, **1997**: p. 46ff.
15 R. Rashed, *Science* **2002**, *297*, 773. This article ascribes an important role in the establishment of experimentation in science to the 10th century Arabic scholar Ibn al-Haytham.
16 B. Gower, **1997**: p. 46.
17 F. Bacon, in *The New Organon and Related Writings,* 1960 ed. (Ed.: F. H. Anderson), Bobbs-Merrill, Indianapolis, IN, **1620**: pp. 19–20.
18 C. G. Hempel, in *The World of Mathematics,* (Ed.: J. R. Newman) Simon and Schuster, New York, **1956.** v. 3, p. 1619ff.

19 C. G. Hempel, in *The World of Mathematics,* (Ed.: J. R. Newman) Simon and Schuster, New York, **1956.** v. 3, p. 1635ff.

20 F. Q. Gouvéa, *Science* **2002**, *296*, 853.

21 F. Bacon, **1620**: p. 115.

22 F. Bacon, **1620**: p. 20.

23 F. Bacon, **1620**: p. 191.

24 F. Bacon, **1620**: p. 96.Aphorism XCIX.

25 **1953**: Concise characterizations of some of Duhem's strengths and weaknesses are to be found in Louis de Broglie's introduction to Duhem's book (**1954**).

26 P. Duhem, **1954**: p. 19.

27 J. H. Poincaré, **1913**: p. 209.

28 A. Ihde, in *The Development of Modern Chemistry,* Harper and Row, New York, **1964.** p. pp. 191–198, 209–216. (reprinted by Dover, New York, 1984)

29 P. Duhem, **1954**: p. 29.

30 A. Ihde, **1964**: Chapter 12.

31 L. Bytautas, D. J. Klein, *Croatica Chem. Acta* **2000**, *73*, 331 and refs. therein.

32 D. J. Klein, L. Bytautas, *Comm. Math. Comp. Chem.* **2000**, 261.

33 J. M. McBride, *J. Am. Chem. Soc.* **1980**, *102*, 4134.

34 P. Duhem, **1954**: p. 299.The translator of Duhem's book points out that this refers to a passage in Plato's Republic, Book VII. In it, Socrates describes the prisoners in the cave as being "unable to turn their heads, for their legs and necks have been fixed in chains from birth. A fire is burning higher up at their backs, and between it and the prisoners there is a road with a low wall built at its side like the screen over which puppet players put up their puppets." Cf. Plato's Republic, edited and translated by I.A. Richards, Cambridge University Press, Cambridge, 1966, p. 123.

35 P. Duhem, **1954**: pp. 38–39.

36 P. Duhem, **1954**: pp. 25–27.Pyrrho was the founder of a school of skeptics in ancient Greece.

37 K. R. Popper, in *The Logic of Scientific Discovery,* K. R. Popper, J. Freed, L. Freed, translators, Routledge, London, **1959.** p. 111. From *Logik der Forschung,* the original German ediition of 1934 (reprinted 1992).

3
Theories from Anywhere: Popper's Philosophy of Conjectures and Refutations

3.1
Popper

Sir Karl Popper (Figure 3.1) has been called by some the greatest philosopher of science of the 20th century. Others, as we shall see, are much less enthusiastic, but those who knew him say he was a formidable character – brilliant, imperious, touchy, and devasting in debate. In a famous book, *The Logic of Scientific Discovery*, first published in 1934 [1] followed by a later companion volume, *Conjectures and Refutations* [2], Popper claimed to have developed by logical analysis a set of criteria for the acceptability of scientific theories and a set of rules for tests of their validity. Fierce debate has swirled around these claims ever since their announcement. To call them "controversial" is an understatement. [3]

Figure 3.1 Karl Raimund Popper. Reproduced from *The Critical Approach to Science and Philosophy* (in honor of Karl R. Popper), (Ed. Mario Bunge). The Free Press of Glencoe, Collier-MacMillan Ltd., London, UK, 1964.

Thus, Popper's ideas are by no means the only or even, arguably, the "best" philosophical framework in which to analyze historical events in chemistry. Nevertheless, despite the disputed nature of the Popperian view of science, I believe that scientists who think about it and explore the philosophical issues raised in opposition to it will find insights into the process of research that can inform their own work and perhaps strengthen it.

3.2 Falsificationism

Popper based his philosophy of science upon what he called his discovery of the essential distinction between a scientific theory and a non-scientific one, a *"demarcation"* between science and metaphysics. From this he developed the idea that progress can be made only by a series of *conjectures and refutations*. According to Popper, an *asymmetry of significance* exists between refutation and verification of a theory. Because no theory ever can be logically *proven* correct, the major work of the experimentalist is to falsify, to prove incorrect, the conjectures of the theorist. This concept is often called "falsificationism," although Popper preferred the term "conjectures and refutations."

3.3 Where Do the Conjectures Come From?

Inductivists like Bacon and Duhem (Chapter 2) visualized the process by which a theory rises to attention to be a systematic development from observations. Experiments lead to generalizations or laws, and underlying physical reasons for these laws are proposed as a higher generalization, a theory.

The falsificationist view presented by Popper dismisses this process as unrealistic and unsuited to real scientific advance. Thus,

> "The theory to be developed in the following pages stands directly opposed to all attempts to operate with the ideas of inductive logic. It might be described as the theory of *the deductive method of testing*, or as the view that a hypothesis can only be empirically *tested* – and only *after* it has been advanced." [4]

Note first that the latter sentence seems to embody the view that there cannot be any significant advance in science by means of accidental or inadvertent discovery. We return to this issue in Chapter 7.

Second, in contrast to the inductivists, Popper apparently does not consider the question of the origin of theories to be of major significance. He consigns this to the psychological realm:

> "If it is the processes involved in the stimulation and release of an inspiration which are to be reconstructed, then I should refuse to take it as the task of the

> logic of knowledge ... It so happens that my arguments in this book are quite independent of this problem ... there is no such thing as a logical method of having new ideas or a logical reconstruction of this process.."

Popper quotes Einstein's remark on the laws of nature:

> "They can only be reached by intuition, based upon something like an intellectual love (*Einfühlung*) of the objects of experience."

Popper's objectives therefore are just to set down the rules of *testing* theories, regardless of where or how these theories originate.

3.4
Falsificationism and the Working Scientist

Of all the philosophers of science, why do I single out Popper for special attention here? The justification runs as follows: The extent to which Popper's ideas have actually guided the work of scientists is difficult to judge, but I have the impression that the force and eloquence of Popper's advocacy, the conviction of his argument, and the apparent simplicity and clarity of his recommendations on procedure have proven to be seductively attractive. Popper is held in special regard by many scientists, who see in falsificationist philosophy a clearly outlined methodological path:

> "Popper speaks to the working scientist in a language that time and again comes straight out of one's heart ... Popper faces on every turn the problems against which the scientist runs up, and solves them ... in full accordance with the procedure of science." [5]

I think a strong case can be made that Popper's ideas exerted an influence on the work of scientists greater than that of any other 20th century philosopher. Of the icons of that field, he was virtually alone in election to Fellowship in the Royal Society, a distinction bestowed on him in 1976. The others who come to mind in this connection and may be found in the Royal Society archives, [6] all had some real scientific credentials: Alfred North Whitehead (elected 1903), Bertrand Russell (1908), and Jules Henri Poincaré (1894) were mathematicians, and Michael Polanyi (1944) was a chemist, so that Popper might be said to be the only pure philosopher so honored. For example, among the luminaries never elected were Ernst Mach, Ludwig Wittgenstein, Thomas Kuhn, and Rudolf Carnap.

Popper developed his scientific philosophy at least in part as a counterweight to the then dominant inductivist fashion. In it, he also opposed the introduction of psychological and even emotional criteria, which had appeared in the influential works of Poincaré and, as we have seen, of Duhem. Here, we examine the nature of some of Popper's ideas. Our aim is to ascertain the reasons for their influence and the actual consequences of attempts to apply them in practice. Our first task will be to determine what Popper himself thought would be the proper relationship between his ideas and the work of scientists.

3.5
Popper's Intent: Descriptive Demarcation or Prescriptive Methodology?

Did Popper intend that his demarcation criterion would strengthen the growth of actual scientific knowledge by providing guidelines for scientific research? Or on the contrary, did he have in mind that his findings were aimed only at providing solutions to internal problems of epistemology?

One author, [7] has expressed the opinion that Popper intended only the latter, that his rules for the logical analysis of scientific theories and experimental results are not to be taken as working guidelines. However, as has been pointed out, [8] the weight of philosophical opinion seems to contradict this view. There can be little doubt that Popper intended his logical deductions to be the basis of a scientific methodology, a set of rules for action. For example, his *Logic of Scientific Discovery* contains a whole section entitled "Methodological Rules." [9] Similarly, that Popper intended his demarcation criterion to be applied in the real world of science seems inescapable from the statement [10]

> "... almost all the problems of traditional epistemology are connected with the problem of the growth of knowledge. I am inclined to say even more: from Plato to Descartes, Leibniz, Kant, Duhem and Poincaré; and from Bacon, Hobbes and Locke to Hume, Mill and Russell, the theory of knowledge was inspired by the hope that it would enable us not only to know more about knowledge, but also to contribute to the advance of knowledge – of scientific knowledge, that is."

Moreover, his intent is clear in passages *inter alia* such as the following from his *Conjectures and Refutations:* [2]

> p. 50 "... I shall try to establish the rules, or if you will the norms, by which the scientist is guided when he is engaged in research or in discovery".
>
> p. 53: "Methodological rules are here regarded as *conventions*. They might be described as the rules of the game of empirical science ...
>
> p. 54, in a section entitled *A Theory of Method:* "Just as chess might be defined by the rules proper to it, so empirical science may be defined by means of its methodological rules. In establishing these rules we may proceed systematically. First a supreme rule is laid down which serves as a kind of norm for deciding upon the remaining rules, and which is thus a rule of a higher type. It is the rule which says that the other rules of scientific procedure must be designed in such a way that they do not protect any statement in science against falsification.
>
> Methodological rules are thus closely connected both with other methodological rules and with our criterion of demarcation [distinguishing science from pseudo-science]. But the connection is not a strictly deductive or logical one. It results, rather, from the fact that the rules are constructed with the aim of ensuring the *applicability* of our criterion of demarcation; thus their formulation and acceptance proceeds according to a practical rule of a higher type." (emphsasis added)

In other words, the falsification criterion of demarcation is not identical to a complete methodological set of rules, but it acts as the "supreme rule" from which the other procedural rules are derived as corollaries.

This view of Popper's intent is widely accepted among philosophers of science, is discussed in many writings on the methodology of science, and forms the basis for much of the philosophical dispute of the last forty years over scientific methodology, [11–18] a debate which continues in both philosophy and science to the present day. [19–35]

Beyond that, one must ask whether Popper's falsification theory of science *requires* that a falsified theory be rejected. The view [7;36] already cited that falsificationism is only a demarcation, not a guide to action, implies that there is no such requirement, that falsification of a theory should not be identified with its overthrow. I find this conclusion unpersuasive. In Lakatos's words [37]

> "... if falsification does not mean rejection, what does it mean? Popper tells us nothing about how we can continue to play the game of science with a falsified hypothesis."

Later chapters of this book give examples of how the game has been played fruitfully even under the aegis of a false theory.

Curiously, one finds a number of statements in Popper's writings that do not mesh with his methodological deductions and actually seem to contradict them. For example, Popper's claim, "my theory of science was not intended to be an historical theory, or to be a theory supported by historical or other empirical facts..." [38;39] seems to be in conflict with all of his recommendations quoted above. How can it possibly be reconciled with his words quoted above "I shall try to establish the *rules ... by which the scientist is guided when he is engaged in research or discovery"* (emphasis added).

If we interpret the italicized phrase to mean this is how science happens, how it has happened, and how it should happen, it is inescapably normative. All of Popper's examples of refutations are taken from historical events. Although he states [39] that "I do not regard methodology as an empirical discipline, to be tested, perhaps, by the facts of the history of science," he also says, regarding the role of his theory of falsification in the history of science, "it plays a leading role ... the examples already given provide some evidence of this, but I shall provide some further examples in the next section." To validate this contention, he follows with a list of twenty examples of key refutations from the field of physics! This seems to me a clear indication that Popper did not intend that his theory stand apart from the sweaty details of scientific events. He intended that scientists use it in their work, and he was proud when they did.

Feyerabend [40;41] gives a dissenting treatment of the question whether one can – or should – make a distinction between "a context of discovery" and "a context of justification." Historical events in science, he contends, occur in "a context of discovery," which may be irrational and need not follow any recognized method. Justification, on the other hand, he says, starts only *after* the discoveries have been made. He quotes from Feigl: [42]

> "It is one thing to retrace the historical origins, the psychological genesis and development, the socio-political-economic conditions for the acceptance or rejection of scientific theories; and it is quite another thing to provide a logical reconstruction of the conceptual structure and of the testing of scientific theories."

Feyerabend expands on this thought:

> "These are indeed two different *things*, especially as they are done by two different *disciplines* (history of science, philosophy of science), which are quite jealous of their independence ... the question is to what extent the distinction drawn reflects a real difference and whether science can advance without a strong interaction between the separated domains ...
>
> ... the attempt to retrace the historical origins, the psychological genesis and development, the socio-political-economic conditions for the acceptance or rejection of scientific theories, far from being irrelevant for the standards of test, actually leads to a criticism of these standards – *provided* the two domains, historical research and discussion of test procedures, are not kept apart by fiat."

The very existence of this book attests that I endorse the value of strong interaction between philosophically derived methodology and historical case studies in science.

3.6
Corroboration vs. Refutation in Experimental Tests of Theory. Popper's Rules

A forward step in science often is represented as the result of a *corroborative* sequence of actions in which a hypothesis is designed to explain observations and then subjected to successive experimental tests of its validity. If the hypothesis survives such tests (unspecified in number or severity), it is said to be valid or accepted. When pressed, we sometimes will admit that such acceptance can only be provisional, but we might as well admit that inwardly, the level of our acquiescence to this uncertainty tends to vary as the inverse of our proprietary interest in the theory under scrutiny.

Thus, experimentalists regard as most admirable those scientific contributions which *corroborate* a theory, especially their own. Yet Popper warns us of a major logical and philosophical problem inherent in this position. Simply expressed, this is a consequence of the fact that no number of corroborations can reduce to zero the possibility that some other theory may in fact be superior. The alternative theory may explain not only all the existing observations apparently confirmatory of the first theory but also may predict new phenomena not predicted by the first and actually *incompatible* with it. This opens the path to a test to determine whether the new prediction can be experimentally confirmed. If so, the second theory becomes preferred, because the first has been *refuted*.

Of course, you will point out immediately that the logico-philosophical situation of the second theory now is indistinguishable from that which the first occupied before its refutation: Who is to say that there cannot be a *third* theory which could supplant the second? But this is precisely the point: all theories that have not yet been actually refuted are nevertheless provisional.

Understandably, the existence of several corroborations of a theory, especially if some of them are spectacular and esthetically pleasing, will create in its proponents a strong psychological bond to it. Eventually, with a large number of independent corroborations, the success of the theory persuades us to accept it provisionally. In fact, science is a vast web of such theories. But, Popper argument seems to tell us, we should not forget that the act of *refutation* of a theory, whether by experiment or by argument, is more decisive, and hence, in a sense, more praiseworthy, than its corroboration.

3.7
The Superiority of Refutations

Thus, according to Popper, [43] the job of the experimentalist is not primarily to confirm theory but rather to refute it. Popper's aim in this formulation is to hold up to skeptical scrutiny the claims of those who would argue that repeated confirmation of a theory constitutes proof of its validity.

The fallacy of proof by mere enumeration is a very old idea in philosophy. As we have seen (Chapter 2), it was well known to Francis Bacon in the 17th century, and to many others, and it has been traced back as far as the writings of Sextus Empiricus, the ancient Greek skeptic. [44] Here is an oversimplified example: In testing the hypothesis "All swans are white," no matter how many white swans we find, we cannot *prove* that no black swans exist, have existed, or will exist, somewhere at some time. Therefore, it is more significant to find a black swan than to find any number of white ones. Refutation rules, in Popper's system. Popper [45] recognizes corroboration of theories as something that may emerge in attempts to falsify a theory, but he accords it a less important role.

Yet many philosophers do not agree. Martin Gardner, [31] a member of the inductivist school and a sometime collaborator with Rudolf Carnap – who was himself a leading exponent of the inductive method and hence a prominent of target of Popper's criticism – gives a caustic summary of some of the counterarguments:

> "Popper's critics insist that 'corroboration' is a form of induction, and Popper has simply sneaked induction in through a back door by giving it a new name ... falsifications are much rarer in science than searches for confirming instances. Astronomers look for signs of water on Mars. They do not think they are making efforts to falsify the conjecture that Mars never had water.
>
> ... To scientists and philosophers outside the Popperian fold, science operates mainly by induction (confirmation), but also and less often by disconfirmation (falsification). Its language is almost always one of induction. If Popper bet on a certain horse to win a race, and the horse won, you would not expect him to shout 'Great! My horse failed to lose!'

... Ernest Nagel ... summed it up this way:

> 'Popper's conception of the role of falsification ... is an oversimplification that is close to being a caricature of scientific procedures.'"

It may be that some philosophers' reluctance to accept Popper's restrictions have to do with the terms "falsification" and "refutation" themselves. The dictionary defines "refute" as "to disprove and overthrow by argument, evidence, or proof; prove to be false or erroneous." Thus defined, the term seems to be derived from formal logic and mathematics. But Popper uses it in a specific technical sense, to refer to what happens in a scientific inquiry. His view of the matter has become gradually less restricted over the years. In fact, one wonders how his falsification theory can be applied at all, when he himself states [46] that

> "no conclusive disproof of a theory can ever be produced ... If you insist on strict proof (or disproof) in the empirical sciences, you will never benefit from experience, and never learn from it how wrong you are."

I conclude from this that what he really means by a scientific refutation is an event or experiment that *casts serious doubt* on a theory. If my surmise is true, we are still left with the problem of just *when* do we abandon a theory. Here he gives us little or no guidance. This problem becomes a major theme in the chapter on Kekulé's theory of benzene later in this book(Chapter 4).

It will be useful to keep in mind the admonitions summarizing Popper's thinking on these issues, which I propose to call "Popper's Rules." These are directly quoted here. [43]

> (1) "It is easy to obtain confirmations, or verifications, for nearly every theory – if we look for confirmations.
>
> (2) Confirmations should count only if they are the result of *risky predictions;* that is to say, if unenlightened by the theory in question, we should have expected an event which was incompatible with the theory – an event which would have refuted the theory.
>
> (3) Every "good" scientific theory is a prohibition: it forbids certain things to happen. The more a theory forbids, the better it is.
>
> (4) A theory which is not refutable by any conceivable event is non-scientific. Irrefutability is not a virtue of of a theory (as people often think) but a vice.
>
> (5) Every genuine test of a theory is an attempt to falsify it, or to refute it. Testability is falsifiability; but there are degrees of testability: some theories are more testable, more exposed to refutation, than others; they take, as it were, greater risks.
>
> (6) Confirming evidence should not count *except when it is the result of a genuine test of the theory;* and this means that it can be presented as a serious but unsuccessful attempt to falsify the theory.
>
> (7) Some genuinely testable theories, when found to be false, are still upheld by their admirers – for example by introducing *ad hoc* some auxiliary assumption, or by re-interpreting the theory *ad hoc* in such a way that it escapes refuta-

tion. Such a procedure is always possible, but it rescues the theory from refutation only at the price of destroying, or at least lowering, its scientific status...

One can sum up all this by saying that the criterion of the scientific status of a theory is its falsifiability, or refutability, or testability."

3.8 Tests of Theory. Refutations and Internal Contradictions

Popper proposed that testing a theory can occur along four different lines: [47] (1) logical comparison of the theory's conclusions among themselves, that is a test for *internal consistency as against contradiction*; (2) a test of logical form to distinguish, for example, a scientific theory from a tautology; (3) a comparison with other theories, with the primary aim of evaluating the theory's power to advance science if it survives the other tests; and (4) tests by empirical applications of the conclusions which can be derived from it, whether by experiment or calculation. We focus our attention on (1) and (4). Much of the discussion over Popper's ideas has centered on empirical tests (point 4), and we shall have more to say about this subsequently, but Popper clearly thought internal contradiction (point 1) to be at least as important.

3.8.1 Logical Inference and Logical Contradiction

Popper set down an absolute prohibition that *a valid theory cannot contain a contradiction*. He considered a self-contradictory theory to be at least as unsatisfactory as one that conflicted with experiment. Just how crucial he considered this to be is best shown in quotation:

> "The requirement of consistency plays a special role among the various requirements which a theoretical system, or an axiomatic system, must satisfy. It can be regarded as the first of the requirements to be satisfied by *every* theoretical system, be it empirical or non-empirical."

Popper makes the following crucial point: The reason we must reject a self-contradictory theory is not just that it is "false." Actually, we frequently work with false statements that, nevertheless, yield results which are adequate for certain purposes. (He gives as an example Nernst's approximation for the equilibrium equation of gases. We might add a more recent case, Hückel's sigma-pi electron separation in chemical quantum theory). But a special kind of falsity is associated with a self-contradictory statement or theory, because a self-contradictory system is *uninformative*. It is so because any conclusion we please can be derived from it. [48–50]

This criticism of self-contradictory theories probably is not familiar to many scientists, but it has been known to philosophers at least as long ago as Duns Scotus, the Scottish scholastic theologician, who died in 1308. It is based upon the laws of logical inference and can be demonstrated rather easily. [48] Thus, from two contradictory statements, any conclusion whatever, whether true or false, may be derived. For example, from the contradictory premises (a) the sun is shining now, and (b) the sun is not shining now, we may deduce any statement, for example, Caesar was a traitor, or 2 + 2 = 5. Popper warns

> "The acceptance of contradictions must lead here as everywhere to the end of criticism, and thus to the collapse of science." [48]

Popper's unequivocally (even dogmatically) emphatic assertion here seems to give the impression that the matter is decided, and further argument is fruitless. I venture to guess that many scientists would agree with him that such a theory must be abandoned. However, we should not conclude that this view is universally accepted among philosophers. Feyerabend, for example, who often can be counted on to challenge Popper's proposals, puts it this way: [51]

> "Finally, the idea that things are well defined and that we do not live in a paradoxical world leads to the standard that our knowledge must be self-consistent. Theories that contain contradictions cannot be part of science. This apparently quite fundamental standard [,] which many philosophers accept as unhesitatingly as Catholics once accepted the dogma of the immaculate conception of the Virgin [,] loses its authority the moment we find that there are facts whose only adequate description is inconsistent and that inconsistent theories may be fruitful and easy to handle while the attempt to make them conform to the demands of consistency creates useless and unwieldy monsters."

Lakatos, [52] Kuhn, [53] Margenau, [54] and Slater, [55] among other authors, have discussed at length one of the most famous examples of an inconsistent – indeed a self-contradictory – theory, namely Bohr's ingenious formulation of his early quantum hypothesis, put forward in 1913, which led to the successful analysis of the atomic spectral lines. The argument need not be repeated here in detail, but the reader may recall that the theory was developed by "grafting" [56] a new set of quantum conditions onto the classical Maxwell-Lorentz theory of electrodynamics. In Margenau's description [54]

> "...it is understandable that, in the excitement over its success, men overlooked a malformation in the theory's architecture; for Bohr's atom sat like a baroque tower upon the Gothic base of classical electrodynamics."

The resulting theory now contained a contradiction, since the classical theory predicted that the emission of light from an excited atom should be continuous, whereas the new quantum conditions allowed emissions only between discrete electronic orbits. This predicted sharp, discrete lines, in agreement with experiment.

Lakatos [56] is much more permissive of this type of contradictory "graft" than is Popper. Lakatos does believe that such inconsistencies ultimately have to be resolved, and he criticizes Bohr's early position that they can and should be tolerated *in principle*. But Lakatos's conclusion resembles that of Feyerabend: the "positive heuristic" that such a flawed theory can make in stimulating further experiment justifies carrying it as long as it continues to have this beneficial effect.

Figure 3.2 Paul Feyerabend. Reproduced from the front dust-cover of *For and Against Method* (ref. 34), courtesy of Dr. Grazia Borrini-Feyerabend.

3.9 Paul Feyerabend and a Stance "Against Method"

Paul Feyerabend's book *Against Method [40]* is perhaps the best known critique of Popper's recommendations. Figure 3.2 may be interpreted to represent Feyerabend's opinion of the Popperian canon.

Apparently, there is a personal aspect to their disagreement, in addition to their philosophical conflict. Popper had become Feyerabend's supervisor when Wittgenstein, [3] who had agreed to allow Feyerabend to join him as a collaborator in Cambridge, died before Feyerabend arrived. Feyerabend credits his appointment to teach philosophy of science at Bristol, despite his lack of formal training in philosophy, to intervention by Popper and Schrödinger (and to "my own big mouth)." [51] Nevertheless, Feyerabend remembers his early acquaintance with a certain reserve:

> "I had met Popper in Alpbach in 1948. I admired his freedom of manners, his cheek, his disrespectful attitude towards the German philosophers who gave the proceedings weight in more senses than one, his sense of humour (yes, the relatively unknown Karl Popper of 1948 was very different from the established Sir Karl of later years) [,] and I also admired his ability to restate ponderous problems in simple and journalistic language. Here was a free mind joyfully putting forth his ideas, unconcerned about the reaction of the 'professionals.'"

But Feyerabend reports that he and other members of his circle of philosophers, even early on, were skeptical about Popper's ideas themselves. He says that deductivism, a major tenet of Popper's thought, was already familiar from Kraft's writing, which predated Popper's. Moreover, the falsificationist philosophy already was taken for granted in the physics seminar of the conference under the chairmanship of Arthur March and

> "so we did not understand what all the fuss was about. 'Philosophy must be in a desperate state,' we said, 'if trivialities such as these can count as major discoveries.'"

Praising Hollitscher, a companion in the Kraft Circle in Vienna, Feyerabend comments that Hollitscher

> "... did not use unfair emotional or intellectual pressures. Of course, he criticized my attitude but our personal relations have not suffered from my reluctance to follow him in every respect. This is why Walter Hollitscher is a teacher while Popper is a mere propagandist."

Apparently, Popper was not the only "disrespectful" philosopher of science! Popper rarely addressed himself to direct rebuttal of Feyerabend's complaints. Anyone who at one time has been a young and unproven scholar will have at least some sympathy for the special anguish of rejection, dismissal, and disregard by a charismatic mentor. One need not approve in order to understand Feyerabend's bitterness. His sharp style here is typical. In fact, most of the criticisms of numerous philosophers and logicians in his book are in the same saucy, iconoclastic, confrontational mode. Aptly, he has been called a "philosophical maverick" [57] with "an urge to raise hackles." [58] His irreverence, his formula for scientific methodology ("anything goes"), his apparent moral obtuseness during the Nazi period in Germany, [22] and his subsequent outspoken radical political views have made him a controversial character. Some of his distinguished targets in the philosophical community find him insufferable, but others find his style refreshing.

The next section outlines some of the main objections to Popper's system in the thoughts of Feyerabend, Duhem, and Kuhn. We shall see that the objections of Feyerabend and Kuhn to rules of procedure center around two not entirely separate concepts: the nature of history and the nature of experimental observation. Duhem also makes a point of the second of these, and to some extent, so does Popper himself.

3.10
"Science is an Essentially Anarchic Enterprise." [59] History Compels Ruthless Opportunism

Feyerabend [59] paraphrases a passage from Lenin [60] and quotations from Butterfield [61] to derive a basis for doubting the efficacy of epistemological rules:

> "History generally, and the history of revolution in particular, is always richer in content, more varied, more many-sided, more lively and subtle than even the best historian and the best methodology can imagine. History is full of 'accidents and conjuncture and curious juxtapositions of events' [61] and it demonstrates to us the 'complexity of human change and the unpredictable character of the ultimate consequences of any given act or decision of men. [61] '"

Quoting Einstein [62] to a similar anti-epistemological effect, Feyerabend expands on this idea:

> "'The external conditions,' writes Einstein, 'which are set for [the scientist] by the facts of experience do not permit him to let himself be too much restricted, in the constructon of his conceptual world, by the adherence to an epistemological system. He, therefore, must appear to the systematic epistomologist as a type of unscrupulous opportunist...'"

The course of scientific research is marked, Feyerabend says, by surprising and unforeseen developments. It demands complex procedures and defies analysis on the basis of rules which have been set up in advance and without regard to the ever-changing conditions of history:

> "... the history of science will be as complex, chaotic, full of mistakes, and entertaining as are the minds of those who invented them. Conversely, a little brainwashing will go a long way in making the history of science duller, simpler, more uniform, more 'objective' and more easily accessible to treatment by strict and unchangeable rules."

3.11
Lakatos, Feyerabend, Kuhn, Hanson: Experiment is Unavoidably Entwined with Theory. The Problem of Incommensurability

One can seek a corollary of these views with the question: What specific aspect of history negates the claim that strict rules can be effective in guiding tests of theory? Reflection leads us to the idea that it is theory itself.

The reigning theory of a scientific field is an organized summary, a distillation, of the history of that field. But the interpretation of experiments designed to test the theory, therefore, will be guided by the theory. If the experiment turns out to be corroborative, there is little motivation to look for other theories that might also explain the results. If the experiment is refutative, a new theory may be required. Thus, what use the investigator makes of the experimental results is determined by theory.

Moreover, theories other than the one under test are unavoidably entwined with the experiment.

Duhem [63] evokes a hypothetical instance in which someone, on a bet, undertakes to describe a physical experiment today by excluding all theoretical language, for example, Regnault's experiments on the compressibility of gases. He would have to rid his account of all the abstract and symbolic expressions introduced by physical theories, that is, the words pressure, temperature, density, weight, optical axis of a lens, etc., and as a result, his account of these experiments alone would fill a whole volume with the most confused, the most involved, and the least comprehensible recital imaginable.

In another passage, [64] Duhem returns to this theme:

"Enter a laboratory; approach the table crowded with an assortment of apparatus, an electric cell, silk-covered copper wire, small cups of mercury, spools of wire, a mirror mounted on an iron bar; the experimenter is inserting into small openings the metal ends of ebony-headed pins; the iron oscillates, and the mirror attached to it throws a luminous band upon a celluloid scale; the forward-backward motion of this luminous spot enables the physicist to observe the minute oscillations of the iron bar. But ask him what he is doing. Will he answer 'I am studying the oscillations of an iron bar which carries a mirror'? No, he will answer that he is measuring the electric resistance of the spools. If you are astonished, if you ask him what his words mean, *what relation they have with the phenomena he has been observing and which you have noted at the same time as he,* he will answer that your question requires a long explanation and that you should take a course in electricity."

3.11.1 Perceptions and Contexts

Hanson [65] paraphrases this story:

"The physicist, in other words, must teach his visitor everything he knows before he can show him what he sees. Not until then will his visitor be supplied with an intellectual context sufficient for throwing into relief those aspects of the cluster of objects before his eyes that the physicist sees as an indication of the electrical resistance of the spools."

Feyerabend's objections to systematic methodology are different in wording but similar in spirit: [66]

"... Not only are facts and theories in constant disharmony, they are never as neatly separated as everyone makes them out to be. Methodological rules speak of 'theories,' 'observations' and 'experimental results' as if these were well-defined objects whose properties are easy to evaluate and which are understood in the same way by all scientists.

However, the material which a scientist *actually* has at his disposal, his laws, his experimental results, his mathematical techniques, his epistemological prejudices, his attitude towards the absurd consequences of the theories which he accepts, is indeterminate in many ways, ambiguous, *and never fully separated from the historical background.* It is contaminated by principles which he does not know and which, if known, would be extremely hard to test. Questionable views on cognition, such as the view that our senses, used in normal circumstances, give reliable information about the world, may invade the observation language itself..."

Feyerabend here is alluding to a much-debated issue in the philosophy of science, which both he [67] and Kuhn [68] have called the problem of *incommensurability*. Kuhn seems to use the word "incommensurable" as a synonym for "theory-laden," whereas Feyerabend uses it in a broader sense, but that difference need not concern us here.

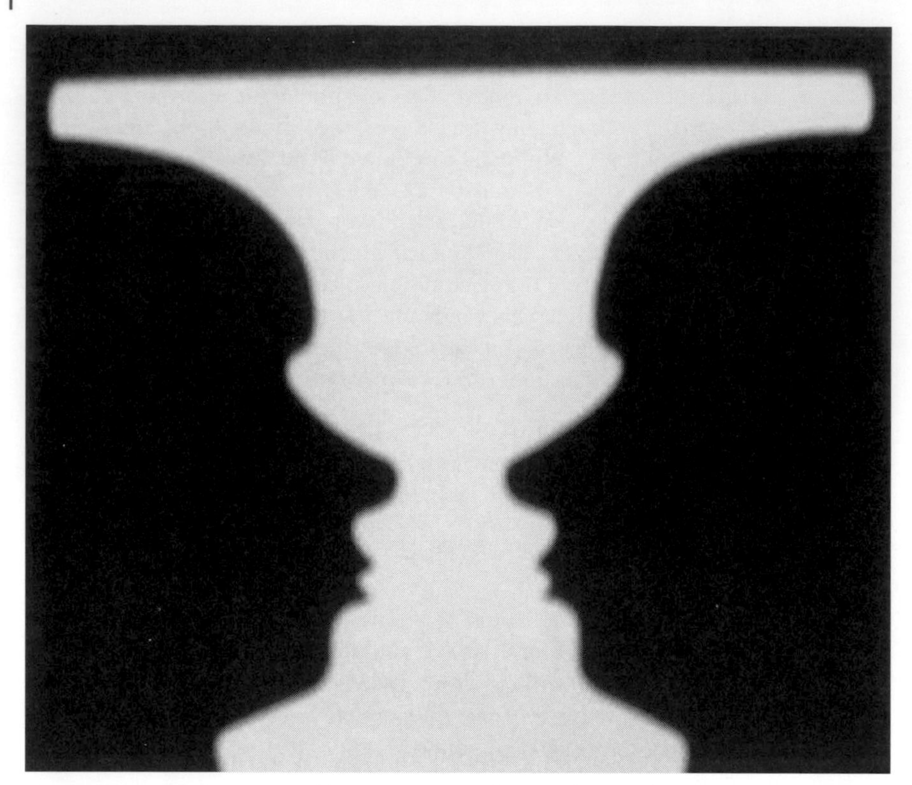

Figure 3.3 Gestalt diagram used by Hanson [69] to illustrate the switch in perception that leads to different interpretations of a given image. **Figures 3.4 and 3.5** are from the same source. Reproduced with permission of W. H. Freeman and Co.

For our purposes, we may say that the concept recognizes that observations made under the influence of an existing theory will take on an entirely new aspect when a new theory replaces the old.

Kuhn draws upon an analogy put forward by Hanson, [65;69] who uses Gestalt diagrams to illustrate this. The demonstration depends on the sudden switch in perception which the viewer experiences when looking at certain images. For example, Figure 3.3 may be seen at first as a silhouette of a Venetian goblet, but in a few minutes, it is seen to be two men staring at each other face to face. Similarly, the shaded portion of Figure 3.4 may at first appear to be a duck looking to the left, but soon it becomes apparent that it can also be a rabbit looking to the right. Note that the the duck image is much easier to see when it is placed in a world of ducks (Figure 3.4), and the rabbit is easier to see in a world of rabbits (Figure 3.5).

Hanson's point, which Kuhn also stresses, is that scientists must make similar adjustments in language, premises, and perceptions that result from a scientific revolution, or as Kuhn would say, a change of paradigm. Adherents of an existing theory will interpret observations in terms of that theory and at first will not have the means

Figure 3.4

Figure 3.5

to understand their significance in a new theory, just as the casual layman visitor to the physics laboratory sees only a light-spot moving across a scale and cannot put the physicist's observations in terms of resistance measurements – or just as the viewer of the duck in the context of ducks is slow to see the rabbit until he or she has accepted the context of rabbits.

3.11.2
Prout's Theory

The story of Prout's theory of 1815 gives a telling example of the operation of incommensurability in science. This is examined historically by Ihde, [70] and philosophically by Lakatos, [52] among others. I briefly summarize the main points here.

In an anonymous article, [71] Prout reported that the densities of gases seemed to be exact multiples of the density of hydrogen. With some diffidence, he suggested that the elements are in fact composed of hydrogen. The hypothesis attracted notable adherents in the immediately following years, but eventually Dumas, and later his student Stas, showed that it certainly did not hold exactly. In a startling example,Stas showed that the atomic weight of chlorine is 35.5 relative to hydrogen as 1, which seemed to be a definitive refutation.

Defenders of Prout's theory tried to save it by suggesting that the analytical samples used for the determinations might contain "impurities" or that the analytical methods might be flawed. These doubts seemed to be put to rest by rigorous methods of sequential fractionation, whose efficacy in producing chemically "pure" samples grew difficult to challenge, and by ever more sophisticated and precise analyses. So Prout's hypothesis slipped into the shadows for a time. Few were left to contest the apparent "fact" that chlorine, for example, a "pure" chemical element, had an atomic weight that deviated significantly from an exact multiple of the atomic weight of hydrogen.

There was, however, a subtle unrecognized assumption in this argument. That was that a "pure" element contains only one kind of atom. In 1871, for example, James Clerk Maxwell [72] thought that there could not be two kinds of hydrogen, "for if some [molecules] were of slightly greater mass than others, we have the means of producing a separation between molecules of different masses. As this cannot be done, we must admit [that all are alike]."

This was all changed by Aston's use of the new method of mass spectrometry to determine atomic weights. His results revealed not only that many elements consist of *mixtures* of atoms of differing atomic weight (isotopes), but at the same time brought to light both the existence and the falsity of the hidden assumption.

Parenthetically, we might note that as subsequent events were to show, there is another reason besides isotopic contamination that the atomic weights of elements are in general not exact multiples of hydrogen. This became apparent only after Einstein's theory of special relativity predicted that in the cosmic furnaces of elemental synthesis, the combination of hydrogens would be accompanied by a relativistic conversion of mass into energy, so that the atomic weights of the higher elements should show a mass defect.

Lakatos points out that not everyone was fooled by the discovery of isotopes. Crookes, in 1886, long before Aston's work, for example, said that

> "Not a few chemists of admitted eminence consider that we have here [in Prout's theory] an expression of the truth, masked by some residual or collateral phenomena which we have not yet succeeded in eliminating"

Lakatos comments

> "That is, there had to be some *further* false hidden assumption in the 'observational' theories on which 'experimental techniques' for chemical purification were based and with the help of which atomic weights were calculated ... Crookes's view even in 1886 [was that] some present atomic weights merely represented a mean value."

Thus, the basic false assumption was

> "... that two pure elements must be separable by *chemical* methods. The idea that two different pure elements may behave identically in all *chemical* reactions but can be separated by *physical* methods, required a change, a 'stretching,' of the concept of 'pure element' which constituted a change – a concept-stretching expansion – of the research programme itself."

This episode makes clear the unavoidable entanglement of our so-called "measurements" and "observations" with theory. One sees this poignantly when one looks again – with hindsight – at the "precise" measurements made during the reign of the theory that each element has one unique mass. In the new theoretical regime, these can look rather foolish. Thus, Soddy [73] gazed sadly upon the frustrated hopes of those who had pursued ever-increasing "precision" in their chemical determinations of atomic weights:

> "There is something surely akin to [,] if not transcending [,] tragedy in the fate that has overtaken the life work of that distinguished gallery of nineteenth-century chemists, rightly revered by their contemporaries as representing the crown and perfection of accurate scientific measurement. Their hard won results, for the moment at least, appear as of little interest and significance as the determination of the average weight of a collection of bottles, some of them full and some of them more or less empty.'"

3.12
Where Do We Go From Here?

Popper's rules thus would impose upon the processes of the scientific enterprise a much more ordered structure than Duhem, Feyerabend, or Kuhn would allow. The confrontation of theories with observed "facts" during the falsificationist test Popper requires seems to call for clean intellectual separation of these two concepts, which if Popper's critics are right, cannot be achieved. These criticisms amount to a severe philosophical challenge to the whole method of conjectures and refutations.

We must be aware that Popper does not accept the criticisms, and that the critics are, on the whole, not even supporters of each others' philosophies. These battles have persisted for decades, a general discord prevails, and the untutored lay chemist, trying to find daylight through the thick undergrowth, may despair of coming to a point of resolution. The absence of consensus among the philosophers leaves us but limited choices: either we must become philosophers ourselves, or we must try to enlighten ourselves by using our common heritage, the accumulated store of historical events in chemistry. There we can find examples to test the philosophers' formulations. This is what I propose to do in some of the following chapters.

The reader may be concerned that I am proposing to examine falsificationism by using falsifying arguments. Is this not a contradiction? We have to remember that falsificationism is not a scientific theory. Among other things, a scientific theory is a generalization intended to explain a set of empirical observations about nature and to predict new ones. To use the term "theory" in describing Popper's rules is potentially misleading. Falsificationism is a recipe for action, a set of instructions on logically acceptable criteria for testing *scientific* theories. It fulfills neither the explicative nor the predictive functions of a scientific theory itself. It is philosophical and metaphysical. It therefore is, potentially and properly, *vulnerable* to oppositional strategies of the type used in debating statements in politics, literature, ethics, religion, or any such body of assertion.

Although one should not rely upon appeals to authority, it is not irrelevant to point out again that neither Popper himself nor any of the many philosophers who have criticized falsificationism have hesitated to use exactly the same kinds of examples as those I give. Popper himself, while vigorously defending falsificationism, has never suggested that falsificatory attacks on it are illegitimate because of this supposed contradiction. In fact, if he believed falsificationism to be a scientific theory, he could hardly make such an objection. That would entangle him in a major contradiction, since he would be saying that falsificationism is immune from falsification, which would violate his requirement that a scientific theory must be, in principle, falsifiable.

Thus, I believe that the justifications of my methods given here permit us to proceed according to plan.

3.13
References and Notes

1 K. R. Popper, *The Logic of Scientific Discovery*, K. R. Popper, J. Freed, L. Freed, translators, Routledge, London, **1959**. From *Logik der Forschung*, the original German ediition of 1934 (reprinted 1992).

2 K. R. Popper, *Conjectures and Refutations: The Growth of Scientific Knowledge*, Routledge and Kegan Paul, London, **1963**. Reprinted from the revised fifth edition of 1989 by Routledge, 1996.

3 D. Edmonds, J. Eidinow, *Wittgenstein's Poker: The Story of a Ten-Minute Argument Between Two Great Philosophers*, Ecco/HarperCollins, New York, NY, **2002**. This book gives the history of a heated philosophical dispute between Wittgenstein and Popper, which took place at Cambridge. Eyewitnesses fifty years later are not unanimous on the details of what happened, but according to one version of

the story, Wittgenstein became so infuriated that he seized a poker from beside the fireplace and brandished it menacingly at Popper. Fortunately, others intervened before serious injury could be inflicted. Chemists also argue, although, I would like to think, usually not at a comparable level of intensity. That is not to say, of course, that the impulse to end the argument by physical means does not lurk in their hearts.

4 K. R. Popper, **1959**: p. 30.
5 H. Bondi, C. W. Kilmister. Quotation on the rear book cover of ref. 1.
6 Raymond and Beverly Sackler Archive of the Royal Society, electronic database, http://www.royalsoc.ac.uk.
7 W. M. Wallau, *Angew. Chem. Intl. Ed. Engl.* **2001**,1201.
8 J. A. Berson, *Angew. Chem. Intl. Ed. Engl.* **2001**,*40,* 1204.
9 K. R. Popper, **1959**: p.78ff.
10 K. R. Popper, **1959**: p. 19.
11 S. Toulmin, *The Philosophy of Science: An Introduction* Harper and Row, New York **1953**. p. 54. p. 201ff. Originally published by Hutchinson & Co. Ltd., London.
12 T. S. Kuhn, *in Criticism and the Growth of Knowledge* (Eds.: I. Lakatos, A. Musgrave), Cambridge University Press, Cambridge, U.K. **1999**. p. 231ff.
13 J. J. C. Smart, *Between Science and Philosophy: An Introduction to the Philosophy of Science,* Random House, New York, **1968**.
14 I. Lakatos, *in Criticism and the Growth of Knowledge* (Eds.: I. Lakatos, A. Musgrave), Cambridge University Press, Cambridge, England **1999a**. p. 91. Reprint of the 1974 revised edition, originally published in 1970.
15 C. G. Hempel, *Philosophy of Natural Science,* E. Beardsley, M. Beardsley, Foundations of Philosophy,Prentice-Hall, Englewood Cliffs, NJ, **1966**. p. 44–45
16 B. Gower, *Scientific Method: An Historical and Philosophical Introduction,* Routledge, London, **1997**.
17 P. Feyerabend, *Against Method: An Outline of an Anarchistic Theory of Knowledge* New Left Books, New York **1975**. p. Reprinted as *Against Method* in 1997 by Verso from the third edition of 1993. p. 150ff.
18 M. Fehér, *Changing Tools: Case Studies in the History of Scientific Methodology* Akadémia Kiadó, Budapest **1995**. p. 56ff.
19 C. Savary, *Phil. Soc. Sci.* **1995**,*25,* 318.
20 D. Papineau, *Times Lit. Suppl.* **1995**,4.
21 D. Papineau, *New York Times Book Review, November 12,* **2000**,28.
22 J. Preston, *Times Lit. Suppl.* **1995**,6.
23 J. Preston, *Phil. Sci.* **1997**,*64,* S241.
24 J. Preston, *Stud. Hist. Phil. Sci.* **1998**,*29,* 425.
25 T. Theocharis, M. Psimopoulos, *Nature* **1987**,*329,* 595.
26 E. M. Kurz, *Creativity Res. J.* **1996**,*9,* 173.
27 J. Watkins, *American Scholar* **1997**,*66,* 295.
28 G. Wächtershäuser, *J. Theor. Biol.* **1997**,*187,* 483.
29 G. Wächtershäuser, *Science* **2002**,*298,* 748.
30 R. Holliday, *BioEssays* **1999**,*21,* 890.
31 M. Gardner, *Skeptical Inquirer* **2001**,*July/August,* 13.
32 D. Miller, *Science* **1999**,*284,* 1625.
33 A. Lazcano, J. L. Bada, *Science* **2002**,*298,* 749.
34 I. Lakatos, P. Feyerabend, *For and Against Method, Including Lakatos's Lectures on Scientific Method and the Lakatos-Feyerabend Correspondence,* University of Chicago Press, Chicago, IL, **1999b**. Edited and with an introduction by M. Motterlini.
35 H. Mortimer, *The Logic of Induction* Ellis Horwood Limited, Chichester, UK **1988**. p. 90ff. With additional material by I. Craig and A.G. Cohn. Translator E. Such-Klimontowicz, translation editors I. Craig and A.G. Cohn.
36 W. M. Wallau, *Angew. Chem. Intl. Ed. Engl.* **2001**,*40,* 1201.
37 I. Lakatos, *in The Philosophy of Karl Popper* (Ed.: P. A. Schilpp), Open Court, La Salle, IL **1974**. v. I p. 241–273.
38 K. R. Popper, W. W. Bartley, III, ed., *Realism and the Aim of Science: From the Postscript to the Logic of Scientific Discovery,*Rowman and Littlefield, Totowa, NJ, **1983**.
39 K. R. Popper, **1983**: p. xxv.
40 P. Feyerabend, *Against Method: An Outline of an Anarchistic Theory of Knowledge,*New Left Books, New York, **1975**. Reprinted as *Against Method* in 1997 by Verso from the third edition of 1993.
41 P. Feyerabend, **1975**: p. 147.

42 H. Feigl, *The Orthodox View of Theories Analyses of Theories and Methods of Physics and Psychology* (Eds.: M. Radner, S. Winokur), University of Minnesota Press, Minneapolis **1970**. p. 4.
43 K. R. Popper, **1963**: p. 36.
44 B. Gower, **1997**: p. 53.
45 K. R. Popper, **1959**: p. 251ff.
46 K. R. Popper, **1959**: p. 50.
47 K. R. Popper, **1959**: p. 32.
48 K. R. Popper, **1963**: p. 317ff.
49 K. R. Popper, **1959**: p. 92.
50 K. R. Popper, **1983**: p.245.
51 P. Feyerabend, **1975**: p. 234ff.
52 I. Lakatos, **1999a**: p. 138ff.
53 T. S. Kuhn, **1999**: p. 256ff. In this paper, Kuhn gives qualified approval to Lakatos's analysis of the events leading up to Bohr's theory.
54 H. Margenau, *The Nature of Physical Reality: A Philosophy of Modern Physics* McGraw-Hill, New York **1950**. p. 307ff.
55 J. C. Slater, *Quantum Theory of Atomic Structure* McGraw-Hill, New York **1960**. p. 1–25.
56 I. Lakatos, **1999a**: p. 187.
57 J. S. Fruton, *Proteins, Enzymes, and Genes* Yale University Press, New Haven, CT **1999**. p. p. 5.
58 *An Urge to Raise Hackles,* N. Maull, New York Times Book Review May 28, 1995 p. 7.
59 P. Feyerabend, *Against Method: An Outline of an Anarchistic Theory of Knowledge* New Left Books, New York **1975**. p. 9.
60 V. I. Lenin, *Left-Wing Communism – An Infantile Disorder, Selected Works,* International Publishers, London, **1967** 3. p. 401.
61 H. Butterfield, *The Whig Interpretation of History,* G. Bell, London, **1931**.
62 P. Schilpp, *Albert Einstein: Philosopher Scientist* Tudor Publishing Co, New York **1957**. p. 683ff.
63 P. Duhem, *The Aim and Structure of Physical Theory,* P. P. Wiener, translator, Princeton University Press, Princeton, **1954**. Reprinted 1991. From the 2nd edition, Marcel Rivière, Paris, 1914.
64 P. Duhem, **1954**: p. 218.
65 N. R. Hanson, *Perception and Discovery: An Introduction to Scientific Inquiry,* (Ed.: W. C. Humphreys), Freeman and Cooper, San Francisco, **1969**.
66 P. Feyerabend, **1975**: p. 50ff.
67 P. Feyerabend, **1975**: p. 219.
68 T. S. Kuhn, *The Structure of Scientific Revolutions,* University of Chicago Press, 1970 ed. p. 198. See also also Sections X, XII.
69 N. R. Hanson, **1969**: Chapters 5–7
70 A. J. Ihde, *The Development of Modern Chemistry* Dover, New York **1964**. p. 497ff. Reprinted in 1983. Originally published in 1964 by Harper and Row.
71 W. Prout, *Ann. Phil.* **1815**,*6,* 321. See also Ann. Phil. **1816**, *7,* 111.
72 J. C. Maxwell, **1871**: As cited by Lakatos **1999a**, p.138ff.
73 F. Soddy, **1932**: As quoted by Lakatos, **1999a**, p. 138ff.

4
Refutation by Internal Contradiction. Kekulé's "Hypothesis of Embarrassment" and the Theory of Benzene

4.1
Internally Contradictory Theories Are Prohibited

Chapter 3 describes Popper's rejection of any internally contradictory theory. Such a theory is unacceptable because it is logically nonsensical: it permits one to derive any statement whatever, true or false. Yet instances exist in which internally contradictory theories are fruitful. A classic case described in Chapter 3 is Bohr's theory of the discrete lines in the spectra of atoms. We now examine a case from organic chemistry, Kekulé's benzene theory, which not only was internally contradictory but which also violated Popper's rejection of *ad hoc* hypotheses.

4.2
Popper Prohibits *Ad Hoc* Rescue of Refuted Theories

Popper set down the condition that once promulgated, a theory could not be unrestrictedly changed in response to experiments that seemed to disagree with it. [1;2] A direct consequence of the falsfication strategy, this prohibition can be seen simply as fair play, for one cannot hope to land a telling blow on a target that will not stand still. He scorned the "strategems," especially the *ad hoc hypotheses,* that theoreticians sometimes use to save a theory that appears to be threatened by a seemingly contradictory finding.

It is necessary to understand Popper's distinction between *auxiliary* hypotheses and *ad hoc* hypotheses. He clarifies this with an example of the apparently anomalous motion of the planet Uranus. Astronomers proposed that this was caused by the presence of a hitherto unknown planet:

> "The new conjecture was auxiliary rather than *ad hoc:* for although there was only this one *ad hoc* reason for introducing it, it was *independently testable:* the position of the new planet (Neptune) was calculated, the planet was discovered optically, and it was found that it fully explained the anomalies of Uranus. Thus, the auxiliary hypothesis stayed within the Newton theoretical framework, and the threatened refutation was transformed into a resounding success.

I call a conjecture *ad hoc* if it is introduced (like this one) to explain a particular difficulty, but (in contrast to this one) *it cannot be tested independently.*"

Feyerabend [3] gives a further example in Newton's attempt to evade difficulties with his theory of light (an illuminating example, so to speak). According to the theory, the surface of a mirror is much rougher than the lateral extension of the light rays. This predicts that a mirror surface should not be reflective. Newton explained this conflict with experiment by means of an *ad hoc* untestable hypothesis:

"...the reflection of a ray is effected, not by a single point of the reflecting body, but by some power of this body which is evenly diffused all over its surface."

This adjustment to the theory made no independent prediction, and hence Popper would consider it *ad hoc.*

Imre Lakatos, [4] a self-proclaimed but not entirely uncritical [5] falsificationist, attempted to defend falsificationism from various objections. He described the conflict between Popper's ideas and those of the so-called "conventionalists" led by Poincaré, whose philosophical system called for adjustments of theory to meet new challenges by the adoption of conventional, agreed-upon hypotheses. Lakatos explained the need for limitations on such hypotheses in the following terms:

"Why aim at falsification at any price? Why not rather impose certain standards on the theoretical adjustments by which one is allowed to save a theory? Indeed, some such standards have been well-known for centuries, and we find them expressed in age-old wisecracks against *ad hoc* explanations, empty prevarications, linguistic tricks."

And in a footnoted passage, he continues:

"Molière, for instance, ridiculed the doctors of his *Malade Imaginaire,* who offered the *virtus dormitiva* of opium as the answer as to why opium produced sleep. One might even argue that Newton's famous dictum *hypotheses non fingo* [I frame no hypotheses] was really directed against *ad hoc* explanations – like his own explanation of gravitational forces by an ether-model in order to meet Cartesian objections."

Or, we might add, like his explanation of why mirrors reflect.

Popper outlines why and how one must avoid *ad hoc* hypotheses:

"Some genuinely testable theories, when found to be false, are still upheld by their admirers – for example by introducing *ad hoc* some auxiliary assumption, or by re-interpreting the theory *ad hoc* in such a way that it escapes refutation. Such a procedure is always possible, but it rescues the theory from refutation only at the price of destroying, or at least lowering, its scientific status..." [6]

"As regards auxiliary hypotheses, we propose to lay down the rule that only those are acceptable whose introduction does not diminish the degree of falsifiability or testability of the system in question, but, on the contrary, increases it....If the degree of falsifiability is increased, then introducing the hypothesis has actually strengthened the theory: the system now rules out more than it did previously: it prohibits more. We can also put it like this. The introduction of an auxiliary hypothesis should always be regarded as an attempt to construct a new system; and this new system should then always be judged on the issue of whether it would, if adopted, constitute a real advance in our knowledge of the world. An example of an unsatisfactory hypothesis would be [one] which had no falsifiable consequences but merely served to restore the agreement between theory and experiment." [2]

Other philosophers swiftly understood Popper's point, as the re-statement by Lakatos exemplifies: [7]

> "...a further condition that a theory must satisfy in order to qualify as scientific: it must predict facts which are *novel*, that is, unexpected in the light of previous knowledge. Thus it is against Popper's code of scientific honour to propose unfalsifiable theories or '*ad hoc*' hypotheses (which imply no *novel* empirical predictions)..."

These strictures imply the imperative that if a theory cannot be saved by any hypothesis other than one that is *ad hoc*, it is falsified and must be abandoned. In Popper's words [8]:

> "...in the case of a scientific theory, our decision depends upon the results of experiments. If these confirm the theory, we may accept it until we find a better one. If they contradict the theory, we reject it."

How should one proceed after a theory has been falsified by Popper's criterion, and hence rejected? Popper has little to offer on this point except the implication that we should try another one. In fact, he says: "In most cases, we have, before falsifying a hypothesis, another one up our sleeves." [9]

Lakatos goes further and asserts that this should not be a matter of choice. Rather, he proposes that "there is no falsification before the emergence of a better theory." [4]

> "If we put forward a theory to resolve a contradiction between a previous theory and a counterexample in such a way that the new theory, instead of offering a content-increasing (scientific) *explanation*, only offers a content-decreasing (linguistic) *reinterpretation*, the contradiction is resolved in a merely semantical, unscientific way. *A given fact is explained scientifically only if a new fact is also explained with it* ... a theory is 'acceptable' or 'scientific' only if it has corroborated excess empirical content over its predecessor (or rival), that is, only if it leads to the discovery of novel facts."

We examine now a famous (and famously fruitful) instance of the rescue of a theory. This example differs formally from Popper's requirement that "...in the case of a scientific theory, our decision depends upon the results of experiments." In the present case, the theory was called into question not directly because of countervailing *experiments*, but rather because it contained a glaring internal contradiction. As Chapter 3 showed, Popper later pointed out formally that any such theory is unacceptable on logical grounds.

But chemists of Kekulé's day, although probably unaware of the detailed reasons why such a theory is philosophically flawed, nevertheless realized that this theory conflicted with Kekulé's own ideas about valence, which had been put forward shortly before. Clearly, some correction had to be found. Kekulé proposed such a correction, but the intellectual construct he put forward for this purpose violated all of Popper's strictures on auxiliary hypotheses.

4.3 Kekulé's Benzene

In 1865–1866, August Kekulé [10–12] (see Figure 4.1) proposed the cyclohexatriene structure for benzene (molecular formula C_6H_6). (I do not intend to participate in the debate [13–17] over whether Loschmidt and other authors may have preceded Kekulé in recognizing the cyclic nature of benzene).

The details of his work and that of other pioneers on the constitution of benzene and related aromatic compounds form a fascinating, intensively reviewed [17–21] chapter in the history of chemistry. He visualized the mental construction of the molecule from an acyclic conjugated six-carbon chain of alternating carbon-carbon single and double bonds, which he represented with the "sausage" formulas in Figure 4.2. [22] The four internal carbons each used three of their valences to form bonds,

Figure 4.1 The Kekulé research group at Ghent in ca. 1860–1865. Kekulé is seated in the center of the front row. Körner stands second from the left in the second row, and Ladenburg stands at the extreme right. Reproduced from R. Anschütz's *August Kekulé*, Band I, Verlag Chemie, 1929, with kind permission of Wiley-VCH.

one single and one double, to adjacent carbons, and the fourth valence to bind a hydrogen. The end carbons were each left with an unoccupied valence (arrow), which was used eventually to bond to the other terminal carbon, forming the ring. Once the ring was formed, all four valences of each carbon were occupied in bonding. In subsequent representations, he omitted the arrows.

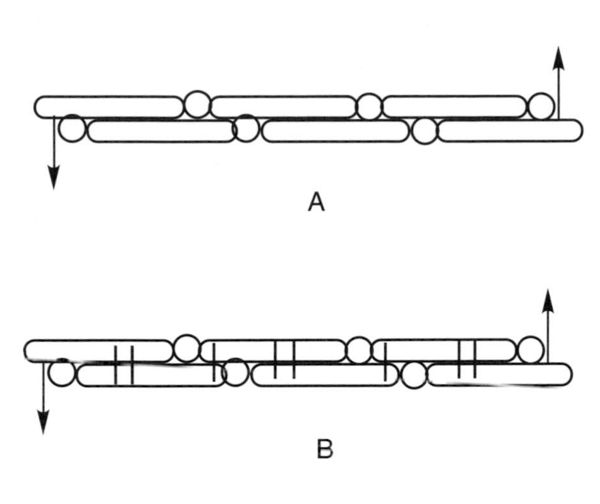

Figure 4.2 A.Kekulé's sausage structure for benzene (*Lehrbuch der Organischen Chemie,* Vol. 2, p. 498). The C-C bonds are formed by contact of portions of each sausage. The arrows represent valences left over at the ends of the chain, connection of which makes the ring-closing bond. Single bonds involve contact of short segments of the sausages, double bonds contact of longer segments of the sausages. See Figure 4.1B. B. Same as A with bonding indicated by one or two vertical lines, corresponding to single or double bonds. Note the pattern of alternating single and double bonds, resulting in the cyclohexa-1,3,5-triene structure.

Kekulé also showed a more graphic representation, which is reproduced here as Figure 4.3. [22] Note the hexagonal arrangement of the carbons and of the hydrogens, and the alternating single and double C-C bonds.

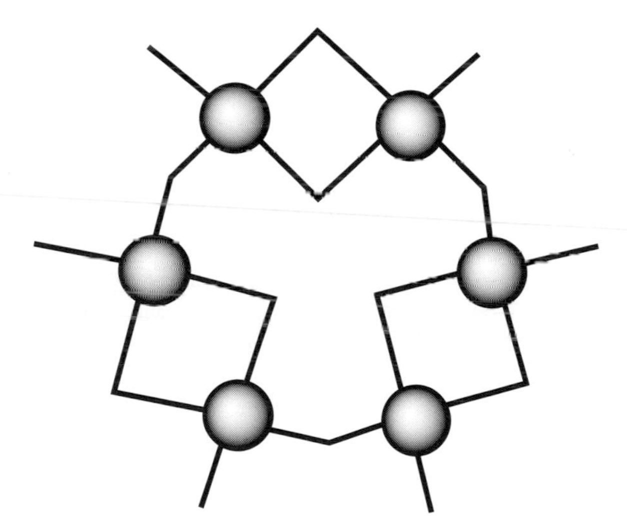

Figure 4.3. Kekulé's graphic cyclohexatriene structure for benzene.

During this period, Kekulé did not use the modern notation we are familiar with (Scheme 4.1, **1a** or **1b**). Rocke has pointed out that, in fact, this was introduced later in 1866 by Claus, who devised this neat representation of Kekulé's idea, published it in an obscure location, and credited it to Kekulé. [17] Claus's representation quickly became standard in the literature and was used by Ladenburg in the 1869 paper (see below) that eventually led to Kekulé's modifying hypothesis.

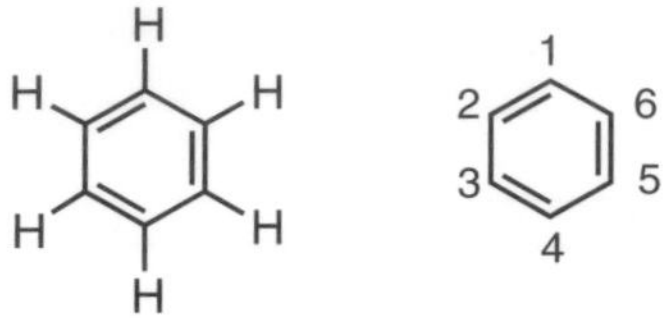

1a 1b

Scheme 4.1

Kekulé's discussion [22] of the 1865–1866 theory thus proposed that benzene is cyclohexatriene. In addition, he made the further important linked stipulations that the molecule is a symmetrical hexagon and that all of the hydrogens are equivalent.

Kekulé's offer of these proposals cleared the ground for a searching test of the theory by the method of isomer numbers:

> "One then easily recognizes that the following isomeric modifications are possible for the increasingly substituted derivatives [Table 1]." [22]

Tab. 4.1 Kekulé's predictions for the number of isomers of substituted benzenes.

Number of hydrogen atoms replaced	*Number of modifications*	*Positions of replaced hydrogen atoms*
one	one	1
two	three	1,2; 1,3; 1,4
three	three	1,2,3; 1,2,4; 1,3,5
four	three	1,2,3,4; 1,2,4,5; 1,3,4,5
five	one	1,2,3,4,5
six	one	–

These predictions cover only the cases where all the substituents are the same. We consider later some cases where they are not.

We must also keep in mind that Kekulé, like all the chemists of that era who used isomer number arguments, was making a *ceteris paribus* (other things being equal) assumption (see Chapter 6), namely that the basic geometry of benzene persists unchanged in its derivatives. For example, it will be obvious from the discussion below that, if the act of substitution were to change the benzene nucleus from regularly hexagonal to a distorted "cyclohexadiene" hexagon in chlorobenzene, and if the distorted structure survived in dichlorobenzene, the number of dichlorobenzenes predicted to be formed would be different from that in Table 4.1.

If position isomers could be found in numbers greater than those predicted by Kekulé's postulates, the theory would be refuted. In Kekulé's words

"If such facts were to be found, then naturally I would have to modify or expand the theory accordingly."

Kekulé's survey of the literature at that time found that there were no authentically confirmed violations of these predictions, although the total number of examples was small. In the succeeding years, he and especially his students Körner and Ladenburg, [23] as well as other chemists, were to perform several isomer number tests of the predictions. Generally, these confirmed the theory. In the next section, we see that nevertheless, there was a profound contradiction at the heart of Kekulé's reasoning.

4.4 Objections from Baeyer, Körner, and Ladenburg

Although the initial reaction of the chemical community to Kekulé's theory was quite favorable, [17] by 1869 a serious criticism had emerged in discussions by Adolf Baeyer, [24] Wilhelm Körner, [25] and Albert Ladenburg, [26] all three of whom were former pupils of Kekulé. Ladenburg's argument appeared in the famous paper [26] in which he proposed the prismatic structure for benzene 2 (Scheme 4.2) as an alternative to Kekulé's hexagonal structure.

Scheme 4.2

2

Although Kekulé had not been explicit about the precise line of reasoning that led to the predictions of Table 4.1, Ladenburg pointed out that he had made a subtle error. [27] Kekulé apparently assumed that if all the hydrogens of benzene are equivalent, then, for example, the two hydrogens ortho to the substituent in a monosubstituted benzene are also equivalent. The opening passage of Ladenburg's paper succinctly summarizes the problem:

"For the clarification of the occurrence of isomers of aromatic compounds, Kekulé proceeds from two hypotheses, which are of the greatest importance for the constitution of benzene and may be expressed as follows:
1. The 6 hydrogen atoms of benzene are equivalent.
2. To every hydrogen of benzene correspond 2 times 2 others which lie symmetrically with respect to it, or if one uses Kekulé's representation, 1.2 = 1.6 and 1.3 = 1.5.
Several years ago I already had the opportunity to point out to Mr. Kekulé that the adjacent graphical formula (**1**, Scheme 4.1) which he had assigned to benzene does not suffice for the hypothesis, since in it 1.2 and 1.6 are not equivalent..."

What Ladenburg is driving at in his rather laconic presentation is that if benzene is a regular-hexagonally symmetrical cyclohexatriene, it is true that all the hydrogens *considered individually* are equivalent, but *considered pairwise*, the pair H_1–H_2 is not equivalent to the pair H_1–H_6. This is because hydrogens 1 and 2 are attached to carbons that are doubly bonded to each other, whereas hydrogens 1 and 6 are attached to carbons that are singly bonded to each other. More generally, one would say today that any two structures which are interconvertible by a symmetry operation are indistinguishable. For example, this is the case for rotation about the C_2 axis which passes through C_1 and C_4 of structure **1a** (Scheme 4.1). This means that H_2 and H_6 are indistinguishable from each other. However, if H_1 and H_2 or H_1 and H_6 are replaced by some group X, there are now two *non-equivalent* structures, as in structures **3** and **4** of Scheme 4.3, which no longer can be interconverted by the C_2 (or any other) symmetry operation. Ladenburg's objection is to Kekulé's implicit and incorrect assumption.

4.5
Tests of Structural Theory by the Number of Isomers

Ladenburg does not spell out all the consequences of this insight in his paper. For example, he discusses the non-equivalence of H_1–H_2 and H_1–H_6, but what about H_1–H_3 and H_1–H_5? Of this, he makes only the cryptic remark that "there can be various opinions." We see from Scheme 4.3 that if we apply the rotational interconvertibility test to the 1,3- and 1,5-disubstituted cases, the molecules are identical when the two substituents are the same, as in 5 and 6, but they are not identical when the substituents are different,as in 7 and 8. In the 1,4-disubstituted case, the substituents (whether X,X or X,Y) do not lead to any extra isomerism.

Scheme 4.3

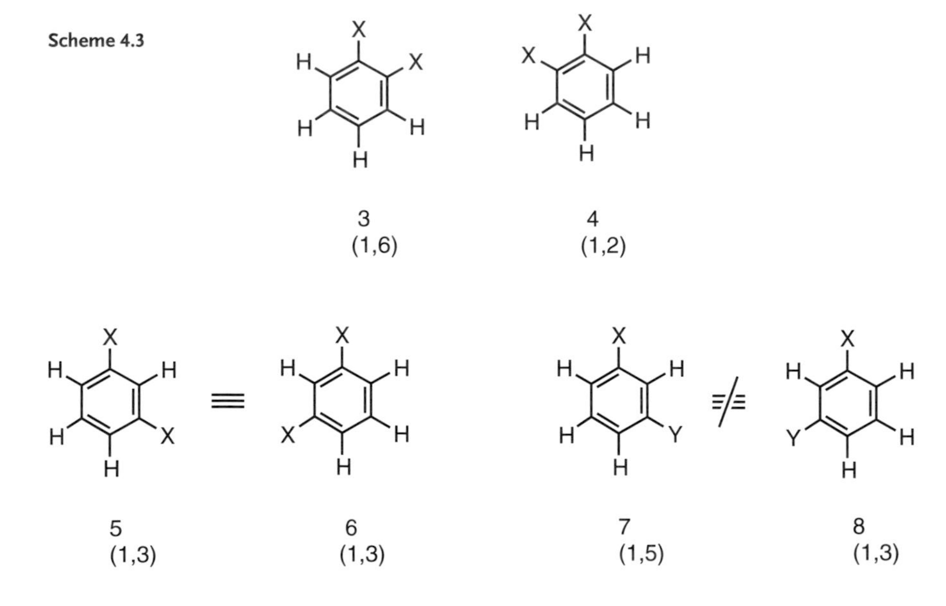

Thus, as Table 4.2 shows, for the class of disubstituted benzenes, there should be a total of 4 isomers when the substituents are the same (2 ortho, 1 meta and 1 para). These predictions (Table 4.2) are quite different from Kekulé's (see Table 4.1). A total of 5 isomers can exist when the substituents are different (2 ortho, 2 meta, and 1 para). Again, these are different from the predictions made from Kekulé's assumption of pairwise equivalence, which would predict only three disubstituted benzenes, regardless of whether the substituents are the same or different.

Tab. 4.2 Number of isomers of disubstituted benzenes predicted by Kekulé's static cyclohexatriene structure for benzene.

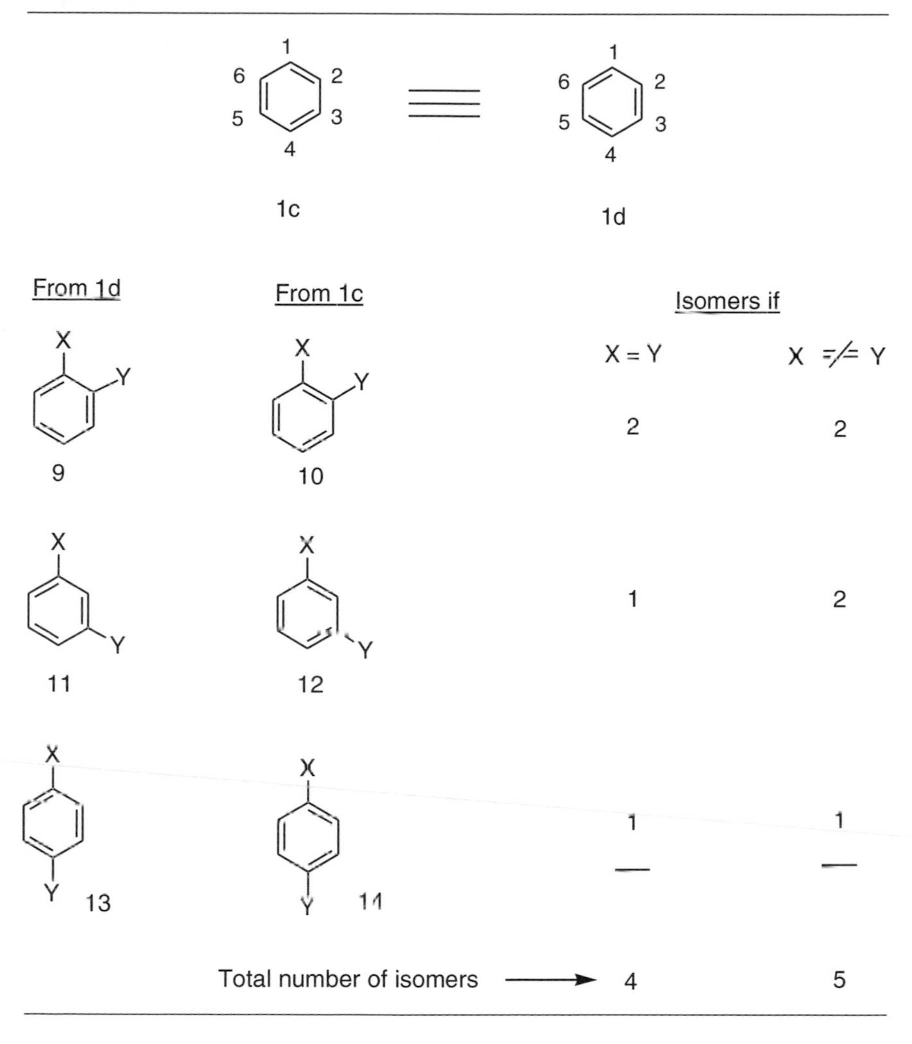

To repeat: The total number of test cases in the literature at this time was small, so that it remained possible that exceptions to the Kekulé isomer number predictions would still be found. However, if this proved not to be the case, chemists would be faced with two difficult problems: (1) Why is it that Kekulé's flawed reasoning with the cyclohexatriene structure apparently leads to correct predictions of isomer numbers for substituted benzenes? (2) Can one imagine alternative structures that would lead to the same predictions but without this ambiguity?

Ladenburg correctly saw that the difficulty with the cyclohexatriene structure was that the nine C-C bonds of cyclohexatriene fell into two groups, three single and three double. He then proposed three structures, **15**, **2**, and **16** (Scheme 4.4) containing only C-C single bonds. At the time, the exact meaning of structure **16**, in which bonds are represented as crossing, was quite uncertain. These can then be considered formal bonding schemes to maintain the quadrivalence of carbon without any multiple bonds. In terms of connectivity, structure **2**, [3]-prismane, represents a normal molecule with carbons clearly tetravalent. It was actually synthesized about a century later. [28;29]

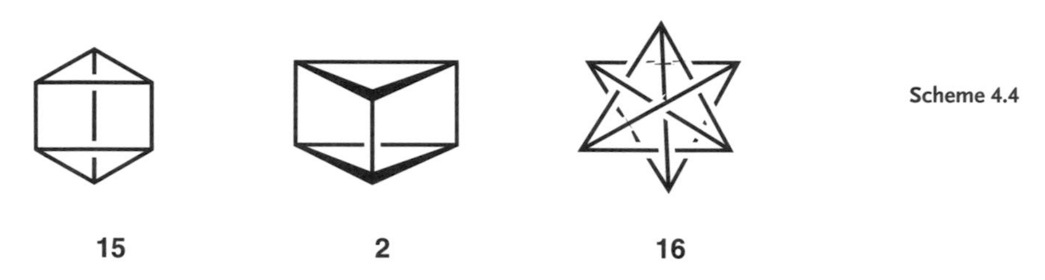

Scheme 4.4

Note that the Ladenburg prismane formula **2** predicts only three isomers of a disubstituted benzene when the substituents are identical (Scheme 4.5, structures **17**, **18**, **19**). One may be tempted to conclude that there is something puzzling here, because two of the structures (**17** and **18**), although not identical, both seem to be "ortho", with substituents on adjacent carbons. However, Ladenburg does not say which of the prismane structures "corresponds" to the one we call "ortho" in the cyclohexatriene structure. Actually, the question is ambiguous, since the connectivities of cyclohexatriene and prismane do not uniquely imply the existence of such a correspondence. His hypothesis says merely that there should be three isomers and only three. [30–32]

Actually, as was soon pointed out by van't Hoff, [33] there is a missing element in Ladenburg's proposal. Ladenburg's prismane structure when substituted as shown with non-identical substituents X, Y, and H, (Scheme 4.5), will lead to two isomers of a type he did not know. These are enantiomers (**20a** and **20b**) rather than structural isomers. The nature of enantiomers as non-superposable mirror images was not to be generally recognized until 1874 with the advent of van't Hoff and LeBel's theory of stereochemistry. Nevertheless, even though the prismatic structure thus had its own difficulties, Ladenburg's objection to Kekulé's structure did not depend upon whether the prismatic structure was correct. He had exposed a major difficulty in Kekulé's cyclohexatriene theory.

Thus, Kekulé's theory consisted of two apparently ill-fitting – in fact, mutually contradictory – parts: (1) benzene is cyclohexatriene; nevertheless, (2) $H_1-H_2=H_1-H_6$, even though literally cyclohexatriene does not allow this. As a consequence, static cy-

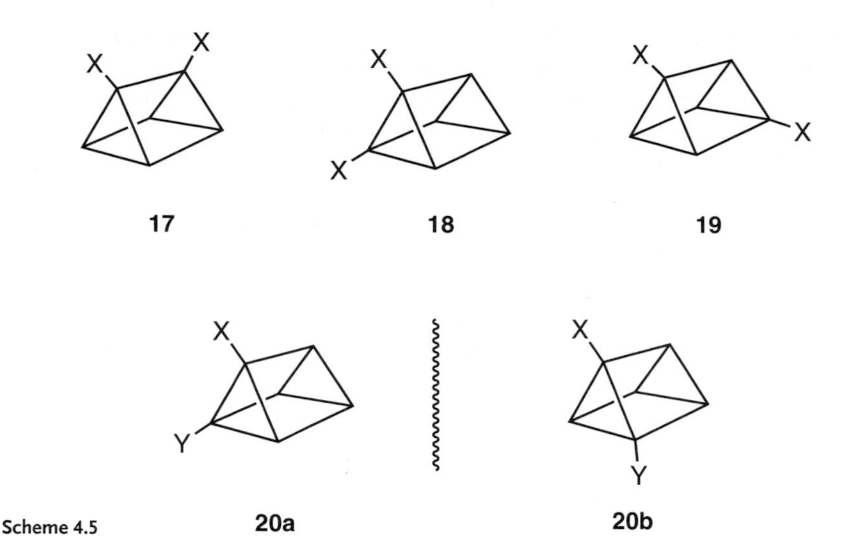

Scheme 4.5

clohexatriene predicts, for example, the existence of two isomeric ortho disubstituted benzenes, but the pairwise equivalence ($H_1 - H_6 = H_1 - H_2$) imposed by Kekulé predicts only one.

Effectively, either Kekulé's benzene theory was wrong, or his theory of valence was wrong. Of course, the idea of cyclohexatriene itself could have survived a refutation based on Kekulé's predictions of isomer numbers in Table 4.1. For example, suppose that two ortho isomers of a disubstituted benzene, one 1,2 and the other 1,6, had actually been found and satisfactorily verified. This is exactly what literal cyclohexatriene requires, but it would have been an outright refutation of the second part of Kekulé's theory. It is not difficult to imagine that such a finding would have been joyfully greeted by the chemical community, which then could forget about the troublesome incompatibility of the two parts together.

Even if no such case of isomerism had been found, a proponent of cyclohexatriene might have argued that the two ortho isomers it predicted might yet be discovered; that is to say, the failure to discover cannot be taken as a demonstration that such isomerism cannot exist. A logical parallel is in the trial lawyer's dictum: absence of proof is not proof of absence.

Similarly, one might have argued that perhaps two ortho isomers made by different routes really are different, but their properties are so similar that they cannot be separated or otherwise distinguished. We might consider this a bit sophistical, but logically it cannot be ruled out.

However, Kekulé declined to use either of these escape routes. He was not interested in a compromise in which he gave up equivalence of pairs of hydrogens in order to retain cyclohexatriene. In fact, the unswerving directionality and force of his presentation leave no doubt that he considered both parts of his theory to be essential. The next section shows the lengths to which he was willing to go in order to save them.

It is not entirely clear why he felt this way. One view held by Walter Hückel [19] seems to suggest that Kekulé simply made a mistake by overlooking the difficulty Ladenburg eventually pointed out. Hückel called Kekulé's *ad hoc* proposal (see the next section) a "hypothesis of embarrassment," which I interpret to mean a rather obvious attempt to salvage the theory by some sort of obfuscation. This suggestion, however, may underestimate Kekulé. I think it is more probable that Kekulé was intuitively convinced that the pattern of substitutions already in the literature reflected reality, rather than mere temporary failure to find the missing isomers. I think he constructed his theory to explain the pattern, and he derived the predictions of Table 1 to make clear his commitment to the theory.

In any case, as a result of the arguments of Baeyer, Körner, and Ladenburg in 1868–1869, Kekulé's theory, if it were to be examined from Popper's viewpoint, either had to be modified in some way, or it had to be abandoned. Popper leaves no doubt about this. In his system, the prohibitions against theories that embody internal contradictions are, if anything, more stringent than those against experimentally refuted theories (see Chapter 3).

4.6 Kekulé's *Ad Hoc* Hypothesis

It took three years for Kekulé to respond to this challenge. [34] He now put forward a completely new and unprecedented hypothesis which significantly modified his theory. This hypothesis was later formulated by others as an oscillation of the bonds in benzene, so that each double bond became single, and each single bond became double. In the case of benzene itself, this would lead simply to an interconversion between two indistinguishable structures **1c** and **1d**, Scheme 4.6). In the ortho-disubstituted case, however, the two tautomeric structures **3** and **4** (Scheme 4.6), *if static,* in principle would be distinguishable, since as we have just seen, they are not interconverted by any symmetry operation. However, if the bond shift is fast on the time scale of observation, structures **3** and **4** would be indistinguishable. This provides the escape route for Kekulé, since now the cyclohexatriene structure would support an *effective* equivalence of C_1–C_2 with C_1–C_6 and thus become compatible with Table 4.1.

It is of great interest, as Gero has pointed out, [21]] that Kekulé's proposal in the new modification of his cyclohexatriene theory was not really an oscillation of bonds. That concept would imply that in half of the benzene molecules the C_1–C_2 bond is single, and in the remaining molecules it is double.

Actually, Kekulé postulated something subtly different. Instead of two rapidly interconverting molecules, he put forward a new theory of the actual bonding forces in benzene. These consist of oscillations of each atom around its equilibrium position, which result in *sequential* (not random) collisions with each of the atoms to which it is bonded. The sequence of collisions is maintained unaltered as time advances. Note that Kekulé proposed this crucial idea without derivation from existing theory, experiment, or other justification. It was completely arbitrary and therefore amounted to an *ad hoc* hypothesis invented to shore up an *ad hoc* hypothesis. In ad-

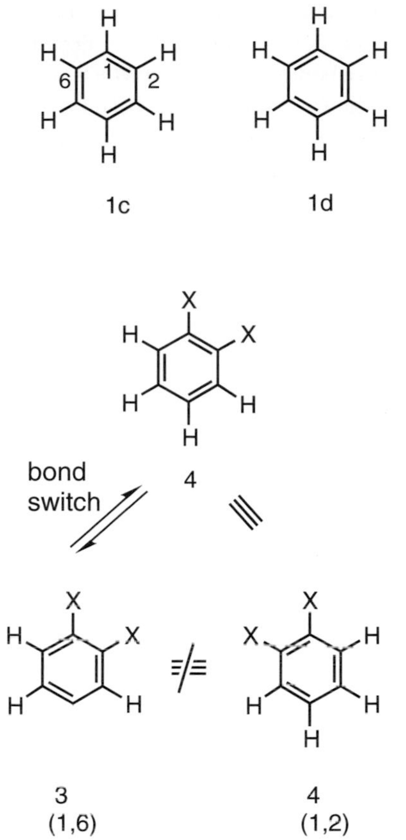

Scheme 4.6

dition, it was a breathtakingly daring (but of course erroneous) hypothesis of a proposed newly recognized property of the valence bond. Now here are the rules of Kekulé's new procedure:

Valence, to Kekulé, is characterized by the frequency of such collisions per unit of time. Thus, the difference between a monovalent and a divalent atom may be exemplified in a diatomic molecule. If the component atoms are both monovalent and are considered to collide once per (arbitrary) time unit, then in a diatomic molecule made up of two *divalent* atoms, the collision frequency in the same time period will be twice that of the monovalent atoms. Further, in a triatomic molecule, such as H-O-H or O=C=O, the middle atom will suffer twice as many collisions in each time unit as the two terminal atoms. By extension, in benzene each carbon will collide with a doubly bonded neighbor carbon twice in a time period, with a singly bonded neighbor carbon once, and with a hydrogen once, so that there will be three collisions with carbons for every collision with hydrogen. Thus C_1 in one of the benzene "structures" **1c** (Table 4.3) collides in the first period with the atoms C_2, C_6, H, and C_2 in that order. In the next period, maintaining the sequence (arbitrarily) assigned, it collides with C_6, H, C_2, and C_6 in that order, corresponding to "structure" **1d**. Over time, C_1 collides *sequentially* with C_2, C_6, H, C_2; C_6, H, C_2, C_6; and so forth. In this sequence, the tetravalence of C_1 is satisfied by the groups of four atoms separated by the semicolons. That is, the second set of collisions corresponds to a bonding pattern in which all the C=C double bonds of **1c** become single, and all the C-C single bonds become double. But it can be seen that whether the molecule is considered to be **1c** or **1d** thus depends on when one starts counting time. Thus, the distinction between **1c** and **1d** is completely arbitrary and without physical meaning. For example, the choice of time period can be eliminated by doubling the time period. Then in each double period, C_2, C_6, H, C_2; C_6, H, C_2, C_6, we see that C_1 collides three times with C_2 and three times with C_6. Thus neither **1c** nor **1d** describes benzene. In the real benzene molecule, all of the carbon-carbon bonds are equivalent, not some single and some double.

Tab. 4.3 Consequences of Kekulé's Collision Hypothesis.

1c **1d**

Collisions of C_1

Time period	*Sequence*	*C_1–C_2 Bond order*	*C_1–C_6 Bond order*
1st, 3rd, 5th..	C_2, C_6, H, C_2 (**1c**)	2	1
2nd, 4th, 6th..	C_6, C_2, H, C_6 (**1d**)	1	2
1st + 2nd	C_2, C_6, H, C_2, C_6, C_2, H, C_6	1.5	1.5
3rd + 4th	ditto	1.5	1.5
5th + 6th	ditto	1.5	1.5

It is worth noting that decades later, Ettore Molinari [35;36] proposed another theory based upon an assumption of sequential collisions of atoms. He called it "motochemie" in German or "motochimica" in Italian, which seems to be translatable as "dynamic chemistry." Although this had elements in common with Kekulé's theory, it was quite different in its predictions. Molinari applied the theory not only to benzene but also to explain several supposed observations by various workers of phenomena that seemed to fall outside the accepted theories of structure and isomerism. Eventually these apparent anomalies were shown to be artifactual, thus guaranteeing a short lifetime for the theory of "motochimie." Despite the theory's virtual disappearance from modern chemistry, it continues to attract attention in the philosophical literature. [37]

We can return to Kekulé's theory with the observation that in an operative sense, the consequences of the two ways of describing benzene, namely Kekulé's original atomic collision concept of valence and the bond oscillation hypothesis into which it later was altered by others, do not differ. Both predict that there should be only one experimentally identifiable ortho-disubstituted benzene. This is true, of course, but the two hypotheses are nevertheless not equivalent. The distinction is that Kekulé's new postulates may be said to provide the *reasons* why the bonds move at all. As we have seen, these reasons are embedded in the conceptualization of valence in Kekulé's formulation.

Kekulé's hypothesis thus was not that the bonds switch, but that nature cannot allow the structure to be a function of the arbitrarily chosen time period in which one decides to examine it. Thus, he says that the question of which form of benzene exists is physically meaningless. Of course, more than a century later, we recognize that his reasoning is tautological: It derives from and is equivalent to his completely arbitrary basic assumptions that the frequency of collisions determines the bond order and that the collisions occur in a fixed sequence.

No physical rationale for Kekulé's assumption of uniform C-C bonds in benzene was available until Erich Hückel's quantum mechanical theories of the 1930s concerning aromatic structure brought forward the idea of delocalization of the π–electrons. Kekulé's collisions of atoms can be considered analogous to the "motions" of electrons in the modern picture. Such "motions" befit a *classical* description but must be modified to take into account the wave nature of the electron in quantum mechanics.

Gero [21] has noted the remarkable resemblance of Kekulé's hypothesis to the more recent concept underlying the so-called "theory of resonance," in which the molecular wave-function is made up of a superposition of contributing "structures" representing the combining wave functions. We might add that, in fact, the principal such contributing structures **1c** and **1d** (Scheme 4.6) are often referred to as "Kekulé structures."

There has been much discussion of whether such an analogy is actually proper. Usually the objection is made nowadays that Kekulé had in mind the tautomerism of two separate molecules, for example, the distorted hexagons **21** and **22** (Scheme 4.7). In a modern view, this would imply that the electronic motion in bond-switching must be accompanied by nuclear motion in order for the short double bonds to become long single bonds, and vice versa. This is not the same as resonance, which demands that the nuclei of one contributing structure remain fixed in the same positions in the other structure.

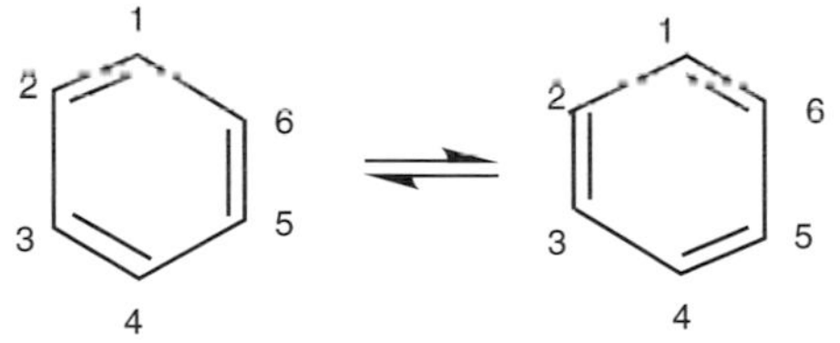

Scheme 4.7 **21** ⇌ 22

However, the objection is dubious, because as we have just seen, such a tautomerism is probably not what Kekulé had in mind. The tautomerism concept was a later interpretation of Kekulé's proposal. At the time, Kekulé had no basis for an opinion on whether single and double bonds should be have different lengths, and he never expressed one. Thus, he could not have been expected to see the full structural implications of the tautomeric interconversion of **1c** $\rightarrow$ **1d**, *even if he had made such a proposal.*

Modern quantum mechanics using the methods of valence bond theory recently has suggested interesting ideas on why the C-C bond lengths of benzene are all equivalent. [38;39]

In Kekulé's time, the only experimental way to determine the hexagonal symmetry of benzene depended upon the method of isomer numbers. As we have seen, this test is most powerful when the number of isomers of a substituted benzene turns out to be *greater* than the number predicted from the symmetrical hexagonal formula, because this may be taken as a refutation. Such techniques as X-ray crystal structure analysis, electron diffraction, vibrational (infrared and Raman) spectroscopy, and nuclear magnetic resonance spectroscopy, which provide direct information about molecular structure, symmetry, and actual bond lengths and convince the modern chemist that the hypothesis of the equivalence of $C_1 - C_2$ and $C_1 - C_6$ in benzene is true, lay decades in the future.

4.7
Violations of Popper's Restrictions

Kekulé's brilliant strategem had no theoretical justification at the time. The collision hypothesis of 1872 itself did not predict any falsifiable consequence then testable beyond what had been in Kekulé's original 1865 proposal. It was only an explicative device for salvaging the cyclohexatriene structure by explaining how "cyclohexatriene" could satisfy the isomer number criterion, not an independently testable hypothesis. Nor did it then (or ever) meet Lakatos's requirement to provide corroborated excess information. Therefore, it clearly violated the Popperian restriction that "only those [auxiliary hypotheses] are acceptable whose introduction does not diminish the degree of falsifiability or testability of the system in question, but, on the contrary, increases it."

Kekulé's justification amounted to an admonition that we should pay no attention to the fact that he wrote double and single bonds in benzene; he argued that for the actual bonds in benzene are not like the familiar single and double bonds of the aliphatic compounds. He wrote them to keep track of the valences but not to represent their actual nature. This property, of course, is not inherent in the original theory of tetravalence of carbon. At that moment, he suggested that the best explanation for this new type of bonding was given by the collision theory.

In sum, it must be said that Kekulé's attempt to patch up the self-contradictory 1865 cyclohexatriene structural theory with the arbitrary and *not then independently testable* collision hypothesis of 1872 violated a cardinal rule of Popper's falsification system.

4.8
Kekulé's Theory: Success Despite Internal Inconsistency

The history of organic chemistry [17–20;40] shows that even though this theory was not really understood by most organic chemists of the 19th century, it was applied nearly everywhere. Chemists of the time quickly suppressed any remaining distaste, swallowed this awkward bolus, and pressed ahead. Their subsequent achievements under the aegis of the theory vindicated their action. The tremendous flowering of synthesis and the discovery of an abundance of new reactions and structures during that time all took place in an atmosphere of growing conviction that, *for whatever reason,* the $C_1 - C_2$ and $C_1 - C_6$ bonds of benzene were structurally equivalent, as Kekulé had said. Even Ladenburg eventually agreed that the Kekulé structure probably was superior to the prism or other alternatives that had been proposed.

Had the chemists of the 19th century obeyed the strictures of Popper and of Lakatos, they should have, at some point, abandoned Kekulé's benzene theory as invalid. Those philosophers' demand for expanded testability of any theory rescued by an *ad hoc* hypothesis does not take into account the historical context of the moment of rescue. How soon should the 19th-century chemists have discarded Kekulé's theory? Suppose we have a theory, like Kekulé's modified, 1872 version, what grounds are we to use to decide whether it is testable other than whether we can think of any way of carrying out such a test? For example, looking forward from our vantage point in the 21st century, we might suggest a test of the modified form of Kekulé's theory, namely the idea that during switching of the double bonds, the electrons of the conjugated system in benzene "move" from a position between two carbons to a position between the next two. It is at least conceivable that such an event could be detected by attosecond time-resolved spectroscopy. However, whether or not such an experiment can ever be done, the fact that one can conceive of such an experiment *now* does not remove the problem of how chemists should have proceeded *in 1872,* when such thoughts were inconceivable.

Remember the major tenets (shown below in italics) of Kekulé's rescuing hypothesis were:

(1) The atoms of a molecule are in constant collision with each other.

Here, I have translated Kekulé's German word "*stoss*" as "collision." Dictionaries translate it alternatively as "recoil" or "impact." The closest we can come to a modern concept resembling Kekulé's collisions is the idea of a molecular vibration.

(2) The collisions occur in a strict sequence.

There is no even remotely comparable modern counterpart to this idea.

(3) The rates of collisions of pairs of atoms with each other are proportional to the number of bonds between them. Thus, C-H = 1, C-C = 1 , C=C = 2, etc.

A present-day test of Kekulé's rescuing hypothesis thus might take the form of determining the *vibrational stretching frequencies* of typical bonds, for example, by infrared spectroscopy, and comparing their ratios to the ratios of the number of bonds. We see immediately that the combination of the modern approximations of Kekulé's hypotheses (1) and (3) is refuted just by inspection of the approximate values observed for the infrared frequencies: C–H 3200 c^{-1}, CC single bond 1450 c^{-1}, CC double bond 1650 c^{-1}, and CC triple bond 2180 cm^{-1}. These values are in the ratio 2.2, 1.0, 1.1, and 1.5, whereas hypothesis (3) requires the ratios 1.0, 1.0, 2.0, and 3.0. Of course, the frequencies and their ratios *are* affected by the number of connecting bonds, but we now know that the frequencies are also profoundly dependent on the *atomic masses*, a fact that Kekulé did not imagine.

More generally, a test that cannot be carried out because either the conceptual understanding or the experimental means to do so are unavailable cannot usefully be postponed for the indefinite future in the hope that such prerequisites eventually will be met. Some old theories, or at least parts of them, tend to lose their meaning in the context of more modern circumstances. If we cannot promptly suggest an experiment to test a hypothesis, it seems to me that Lakatos's criterion of a "positive heuristic," is a sensible way to proceed. As long as the theory embodying the hypothesis continues to generate correct predictions, continued use of it is justified. In fact, Lakatos points out [7] that this is just what often happens in the history of scientific research. Despite Popper's objections, researchers tend to find excuses for not abandoning a functioning theory too quickly.

4.9 Falsifiability in Principle

Popper [41]has attempted to forestall or deflect criticism [3;5;7;42;43] of the falsification methodology by drawing a distinction between the philosophical logical-technical concept of "falsifiable" and the term "demonstrably falsifiable." Thus, in regard to this distinction, he writes

> "Although the the first sense refers to the logical possibility of a falsification in principle, the second sense refers to *conclusive practical experimental proof* of falsity [emphasis in the original]."

In this spirit, a determined rear-guard action in response to the present example of Kekulé's theory of equivalent C-C bonds might be that nothing in Popper's philosphy requires that we reject the theory, since falsifiability of the rescuing collision hypothesis still was possible *in principle*. One cannot judge the working chemist harshly for being skeptical about the utility of such an escape and for moving on without pondering the issue. How long are scientists supposed to wait for the required corroboration of a (not-yet-articulated) independent prediction by the rescuing hypothesis to appear? It seems to me that, aside from theories excludable on logical grounds such as internal contradiction, most (if not all) theories *eventually may be* falsifiable (and hence testable) in principle. If we are trying to decide whether to use and apply a proposed rescuing theory, the important question is not whether it is testable in principle, but rather whether an independent test of it can be conceived. This implies that the new test should be suggested immediately or very soon. Otherwise, scientists are justified in looking elsewhere for a better theory, or in continuing to use the "flawed" theory, despite its violation of falsificationist principles.

What made Kekulé's collision hypothesis unsatisfactory in Popper's system was not that the hypothesis is wrong, although that is so in terms of modern theory, but rather that in Kekulé's time and for decades afterwards, neither such a test nor any other direct test of the components of the rescuing hypothesis was possible.

In fact, it could be argued that the theory's *testability in principle* could not have been judged at the time the theory was proposed. Testability in principle was an evaluation that could have been made only retrospectively. I do not suggest that Kekulé's rescuing hypothesis cannot ever be tested. But the real pragmatic question is what are we to do with a "flawed" theory in the meantime?

4.10 Reflections

In our daily encounters with nature, we are aware of all kinds of theories, some tested by experiment, some logically testable but not yet tested, *and some not testable by any procedure known or conceivable at the time*. Thus, whether a theory is testable in practice, using available or readily derivable methods, or whether no method currently exists and the theory is therefore testable only in principle, our decision on whether to continue to use the theory must be based upon pragmatic considerations. Even for a "refuted" theory, we must still ask: Does it predict new phenomena or rationalize a body of information? In Lakatos's words [44]

> "We may appraise research programmes, even after their 'elimination', for their *heuristic power:* how many new facts did they produce, how great was their capacity to explain their refutations in the course of their growth?"

The 19th century organic chemists were able to create the sturdy foundation of their subject while relying upon Kekulé's logically "flawed" theory, which embodied an "unacceptable" internal contradiction, and which he attempted to rescue with an "unacceptable" *ad hoc* modification. What path of development might organic chemistry have taken if instead the chemists of that time had adopted a falsificationist stance and rejected Kekulé's theory outright? Would they have waited sixty years until the quantum mechanical theories of the 1930s eventually made Kekulé's speculations respectable? Obviously no answer other than pure conjecture can be given now, but the facts in this case, as in many others, argue that working scientists charged ahead, even in violation of strict logic and rationality. In the end, the process by which Kekulé's theory became accepted seems to have been a series of experimental steps, tentative at first and then ever more confident, each of which affirmed the theory, whatever its flaws.

4.11 Acknowledgement

This chapter is an expanded version of the article *Kekulé Escapes, Popper Notwithstanding*, which appeared as *Angew. Chem. Intl. Ed. Engl.* **2000**, *39*, 3045; *Angew. Chem.* **2000**, *112*, 3173.

4.12 References and Notes

1 K. R. Popper, *The Logic of Scientific Discovery*, K. R. Popper, J. Freed, L. Freed, Routledge, London, **1959**. Translated from *Logik der Forschung*, the original German ediition of 1934 (reprinted 1992), by K.R.Popper, J. Freed, and L. Freed.

2 K. R. Popper, **1959**: p. 82ff.

3 P. Feyerabend, *Against Method* Verso (New Left Books), New York **1975**. p. 44ff. (Reprinted in 1997 from the 3rd ed. of 1993, 1st ed. 1975).

4 I. Lakatos, *Criticism and the Growth of Knowledge: Proceedings of the International Colloquium in the Philosophy of Science, London, 1965* (Eds.: I. Lakatos, A. Musgrave), Cambridge University Press, Cambridge, UK **1999**. p. 117. Original publication in 1970, reprinted in 1999.

5 I. Lakatos, *The Philosophy of Karl Popper* (Ed.: P. A. Schilpp), Open Court, La Salle, IL **1974**.v. 1 p. 241–273.

6 K. R. Popper, *Conjectures and Refutations: The Growth of Scientific Knowledge* Routledge and Kegan Paul, London **1963**. p. 37. (Reprinted from the revised fifth edition of 1989 by Routledge, 1996).

7 I. Lakatos, *in Boston Studies in Philosophy of Science* (Eds.: R. Buck, R. S. Cohen), Reidel, Dordrecht **1971**.v. 8 p. 91ff.

8 K. R. Popper, *The Open Society and Its Enemies* Routledge, London **1945**.v. 2 p. 233.

9 K. R. Popper, **1959**: p. 87.

10 A. Kekulé, *Lehrbuch der Organischen Chemie* Ferdinand Enke, Erlangen **1866**. v. 2 p. 494–515.

11 A. Kekulé, *Ann.* **1866**, *137*, 158.

12 A. Kekulé, *Bull. Soc. Chim.* **1865**, *[2] 3*, 98.

13 *The Kekulé Riddle: A Challenge for Chemists and Psychologists*, 1st Cache River Press, Clearwater, FL, **1993**. J. Wotiz, ed.

14 A. Bader, *Richard Anschütz, Archibald Scott Couper, and Josef Loschmidt Pioneering Ideas in Physical and Chemical Science, Proceedings of the Josef Loschmidt Symposium* (Eds.: W. Fleischacker, T. Schoenfeld), Plenum, New York, NY **1997**. p. 101. Meeting date 1995.

15 C. Noe, A. Bader, *Chem. Brit.* **1993**, *29*, 126.

16 J. Wotiz, S. Rudofsky, *Chem. Brit.* **1984**, *20*, 720.

17 A. J. Rocke, *Ann. Sci.* **1985**, *42*, 355 and references cited therein.

18 S. G. Brush, *Stud. Hist. Phil. Sci.* **1999**, *30*, 21 and references cited therein.

19 W. Hückel, *Theoretical Principles of Organic Chemistry* Elsevier, Amsterdam **1954**. v. 1 p. 642ff.

20 A. J. Ihde, *The Development of Modern Chemistry* Dover, New York **1984**. p. 310–316. (reprinted from the 3rd edition of 1970 published by Harper and Row).

21 A. Gero, *J. Chem. Ed.* **1954**, *31*, 201.

22 A. Kekulé, *Lehrbuch der Organischen Chemie* Ferdinand Enke, Erlangen **1866**. v. 2 p. 515.

23 K. Hafner, *August Kekulé dem Baumeister der Chemie zum 150. Geburtstag*, Justus von Liebig Verlag, Darmstadt, **1980**.

24 A. Baeyer, *Zeitschr. Chem.* **1868**, *4 (Neue Folge)*, 120.

25 W. Körner, *Giornal. di Scienze Natur. ed Econ.* **1869**, *5*, 212.

26 A. Ladenburg, *Ber.* **1869**, *2*, 140.

27 Baeyer seems to have been the first to publish an explicit objection: "According to Kekulé's benzene formula, one can insert three like elements into benzene in four different ways, since, for example, two neighboring carbon atoms each bearing a substituent can be bound to each other in two non-equivalent ways, that is, by either a single or a double bond."

("Man kann nämlich nach der Kekulé'schen Benzolformel drei gleichartige Elemente auf 4 verschiedene Weisen in das Benzol einfügen, wenn man berücksichtigt, dass es nicht gleichgültig ist ob z.B. 2 benachbarte mit einer Seitengruppe verbundenen C-Atome mit ein oder mit 2 Affinitäten mit einander verbunden sind.") The papers of Körner and Ladenburg both appeared in the following year 1869, and the wordings of the crucial point regarding the pairwise non-equivalence of C_1-C_2 and C_1-C_6 in their papers is almost identical in the two. Although Molinari (E. Molinari, *J. prakt. Chem.*, **1893**, *48*, 113) implies that Ladenburg was a later participant, this does not agree with Ladenburg's claim that he spoke to Kekulé about the issue of the number of isomers "schon vor mehreren Jahren." In the context of the discussion in this chapter, the more important point to make is that the internal contradiction in Kekulé's theory already was recognized by a substantial group of leading investigators only a short time after the theory had been promulgated.

28 T. J. Katz, N. Acton, *J. Am. Chem. Soc.* **1976**, *95*, 2738.

29 H. Hopf, *Classics in Hydrocarbon Chemistry* Wiley-VCH, Weinheim **2000**. p. 42ff. A review.

30 D. J. Klein, *Comm. Math. Computer Chem.* **2000**, 261. This paper gives a more general graph theoretical procedure for enumeration of isomers. See also the following two references.

31 D. J. Klein, L. Bytautas, *Comm. Math. Comp. Chem.* **2000**, 261.

32 J. M. McBride, *J. Am. Chem. Soc.* **1980**, *102*, 4134.

33 J. H. van't Hoff, *Ber.* **1876**, *9*, 1881.

34 A. Kekulé, *Ann.* **1872**, *162*, 77.

35 E. Molinari, *J. prakt. Chemie* **1893**, *[2]*, *48*, 113.

36 E. Molinari, *Gazz. Chim. Ital.* **1893**, *23*, 47.

37 N. Psarros, *Die Chemie und ihre Methoden*, Wiley-VCH, Weinheim, Germany **1999**.

38 A. Shurki, S. Shaik, *Angew. Chem. Intl. Ed. Engl.* **1997**, *36*, 2205.

39 P. Hiberty, D. Danovich, A. Shurki, S. Shaik, *J. Am. Chem. Soc.* **1995**, *117*, 7760.

40 A. J. Rocke, *The Quiet Revolution: Hermann Kolbe and the Science of Organic Chemistry* University of California Press, Berkeley **1993**. p. 296–304.

41 K. R. Popper, *Realism and the Aim of Science, from the Postscript to the Logic of Scientific Discovery* (Ed.: W. W. Bartley), Rowman and Littlefield, Totowa, New Jersey **1956, 1983**. p. xxxi–xxxv.

42 N. R. Hanson, *Perception and Discovery* (Ed.: W. C. Humphreys), Freeman and Cooper, San Francisco **1969**. p. 278–279.

43 T. Kuhn, *Logic of Discovery or Psychology of Research? The Philosophy of Karl Popper* (Ed.: P. A. Schilpp), Open Court, La Salle, IL **1974**. v. 2 p. 798–819.

44 I. Lakatos, **1999**: p. 137.

5
Are Crucial Experiments Conclusive? Refutations: Permanent, Temporary, and Virtual

5.1
Refutations and "Crucial" Experiments

To have meaning in Popper's scheme, the word "refutation" must have an implication of finality. Refutations, according to Popper, require a "crucial" experiment, which he defines as "one which is designed to bring about a decision between two competing theories by refuting (at least) one of them – without, of course, proving the other." [1] That is, at some point, one has to decide when a theory actually has been refuted. Otherwise, Popper's prohibition of the rescue of refuted theories by *ad hoc* hypotheses carries no force. Of course, the refuter may *claim* that he has carried out a "crucial" experiment and hence that the result must be taken as final, but in practice, this is not always the case: some refutations gain acceptance only slowly. The operative criterion of finality is acceptance of the refutation by consensus of members of the community of scholars working in the field. This may be displayed by explicit statements in the literature, or more convincingly, by the cessation, sometimes immediately but often gradually, of research efforts based upon the discredited theory. The process has been described by, among others, Kuhn [2] and Galison. [3]

However, even this may be insufficient to deal the coup-de-grace. Sometimes the supposed "refutation" is not final because the theory, although one of its predictions is incorrect, nevertheless retains a range of validity. Not infrequently, even an "incorrect" prediction is subsequently seen to have been irrelevant rather than truly erroneous, because it was based upon an inappropriate definition of the field of applicability of the theory. Beyond that, there are cases in which a theory, after having been refuted in a way that is convincing to the great majority of workers in a field, and after having been abandoned by its own author as false, nevertheless is resurrected by subsequent events. We study now some cases of refutations, actual or presumed, which exemplify the complexities and uncertainties of the process. We shall find theories that spring back to life after having been buried, as well as theories that died even though their last remnants of vitality had not yet been stamped out.

5.2
Death and Resurrection of Theories. Temporary Refutations

5.2.1
Meerwein's Hypothesis of Non-Vicinal Hydrogen Shift in the Racemization of Camphene

In their memorable work on the rearrangements of bornyl and isobornyl derivatives of camphor **1** (Scheme 5.1), Meerwein and van Emster [4;5] took up again an old problem: It had been known for some time that the dehydration of borneol **2** or isoborneol **4** under acidic conditions, and the dehydrohalogenation of bornyl chloride **3** or isobornyl chloride **5** often led to partial or complete racemization of the product camphene **6**. That is, for example, a sample of isobornyl chloride **5** consisting of *only one* of its two non-superposable mirror-image forms gave camphene **6** consisting of a *mixture* of its two non-superposable forms.

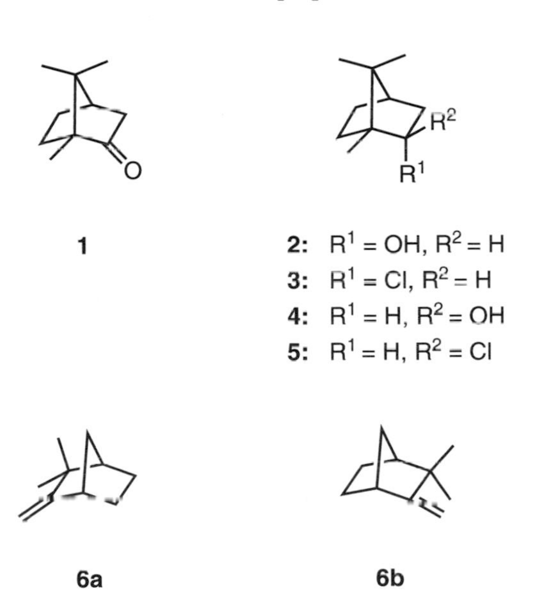

Scheme 5.1

Meerwein and van Emster correctly surmised that the partial racemization had great mechanistic significance. They showed that the degree of racemization depended strongly on the experimental conditions, increasing with an increase in temperature, time, or acid concentration. They also found that isobornyl chloride **5**, recovered from contact with a Lewis acid catalyst such as stannic chloride, was itself partially or fully racemized. It will be instructive to lay out the intricate story of how Meerwein was led by his own misinterpretation of a serendipitous finding in another system to an apparently incorrect theory of the camphene racemization, how he was severely criticized for both the error and the derived theory, and how nevertheless, the theory ultimately proved to be valid.

Organic chemists will recall that the key element of Meerwein's interpretation of the Wagner rearrangements was the migration of a carbon (C_6) with its attached substituents from C_1 to C_2 of a bicyclic cationic intermediate (**7** → **8**, Scheme 5.2), followed by succesive formation and dehydrohalogenation of camphene hydrochloride **9**.

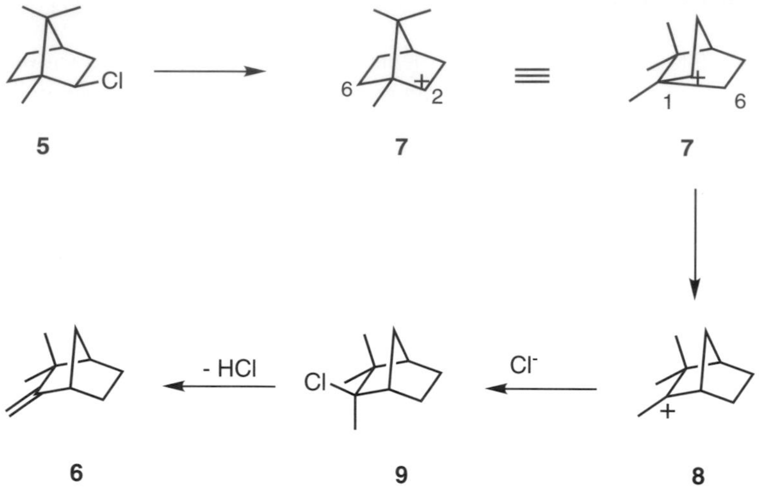

Scheme 5.2

This mechanism accounted nicely for most of the experimental facts, in particular for the catalysis of the reaction by Lewis acids and for the increased rate of the structural rearrangement in polar, ionizing solvents. There was, however, one bothersome detail: Scheme 5.2 did not provide a racemizing pathway. Therefore, the camphene obtained from enantiomerically pure isobornyl chloride **5** should also have been enantiomerically pure. But as we have just seen, in general it was not. Similarly, no mechanism is apparent from Scheme 5.2 for the self-racemization of isobornyl chloride. Obviously, something has been left out of Scheme 5.2 as it stands.

Now here is where Meerwein got into trouble. His attempt to explain the racemization received what he interpreted to be an encouraging stimulus from a study of a different molecule, reported with Wortmann in 1924 [6] soon after the van Emster papers [4;5] of 1920–1922. For reasons that need not concern us here, Meerwein and Wortmann studied the rearrangement of "α-camphor dichloride" (more properly called 2,2-dichlorocamphane) **10** (Scheme 5.3), obtained by the action of phosphorus pentachloride (PCl_5) on camphor **1**. Upon treatment of α–camphordichloride with stannic chloride, they observed the formation of a new substance isomeric with the α–isomer, which they named β–camphordichloride.

The structure they proposed for β–camphordichloride was 2-*endo*-6-*exo*-dichlorocamphane **11** (Scheme 5.3). The reader may be spared a detailed account of their reasoning, which was convoluted, in part self-contradictory, and finally just wrong. There were, however, two key sets of reactions that deserve further discussion.

The first is the conversion of each of the isomers, in phenolic solution containing potassium acetate, to a dehydrohalogenation product, the α–isomer to α–chloro-

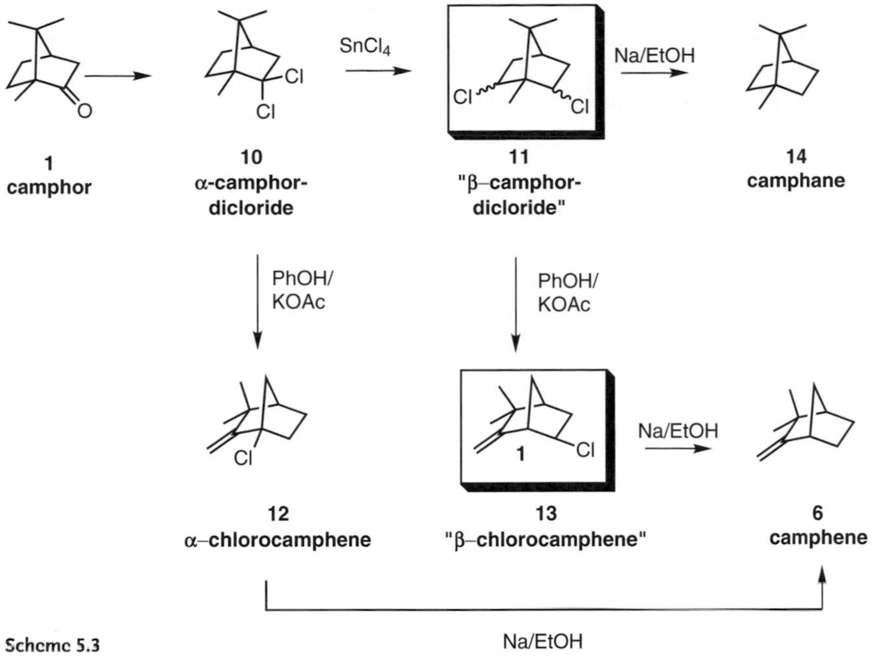

$SnCl_4$ Na/EtOH

1 camphor
10 α-camphor-dicloride
11 "β-camphor-dicloride"
14 camphane

PhOH/KOAc PhOH/KOAc

Na/EtOH

12 α–chlorocamphene
13 "β–chlorocamphene"
6 camphene

Na/EtOH

Scheme 5.3

camphene, which Meerwein and Wortmann [6] formulated as **12**, and the β to β–chlorocamphene, formulated as **13** (Scheme 5.3). The α– and β–chlorocamphenes differed markedly in physical properties, but both gave camphene itself **6** upon reduction with sodium and ethanol. Scheme 5.3 gives the structures as assigned by those workers [6] but two of the structures, **11** and **13**, were later found to be incorrect (see below) and are shown here in frames.

The second reaction was the reductive dehalogenation of β–camphordichloride to the known saturated hydrocarbon camphane **14**. Since **14** embodies the same carbon structure as the starting α camphordichloride (**10**, 2,2-dichlorocamphane), Meerwein and Wortmann incorrectly concluded that "therefore an alteration of the carbon skeleton in the conversion of α– to β–camphordichloride has not occurred." The reader will recognize that the meaning of this conclusion depends on how the term "alteration of the carbon skeleton" is interpreted. As we shall soon see, it would be possible to imagine, for example, that such a rearrangement has occurred, but that the apparently unaltered skeletal structure does not reveal it. It appears to me that this error in interpretation was the main reason for the assignment of the structure **11** to β–camphordichloride. Meerwein and Wortmann noted that **11** could be formed without carbon shift by shifts of a hydrogen from C_6 to C_2 and a chlorine from C_2 to C_6 of the bicyclic skeleton, a process that had no recognized precedent in the literature.

Meerwein was dazzled by the implications of the new theory, which was based upon this supposed discovery. He and Wortmann [6] proposed that the 6,2-shift was the explanation for the rearrangement of fenchol **15** to isofenchol **16** (Scheme 5.4), and in a paper published in 1924 together with the Wortmann paper, Meerwein and Montfort [7] recognized that it would interconvert the enantiomeric isobornyl cations **7a** and **7b**, thereby providing, they thought, the long-sought solution to the riddle of the racemization in the rearrangement of isobornyl chloride and isoborneol to camphene (Scheme 5.5).

6 2 OH HO

15 **fenchol** **16** **isofenchol**

Scheme 5.4

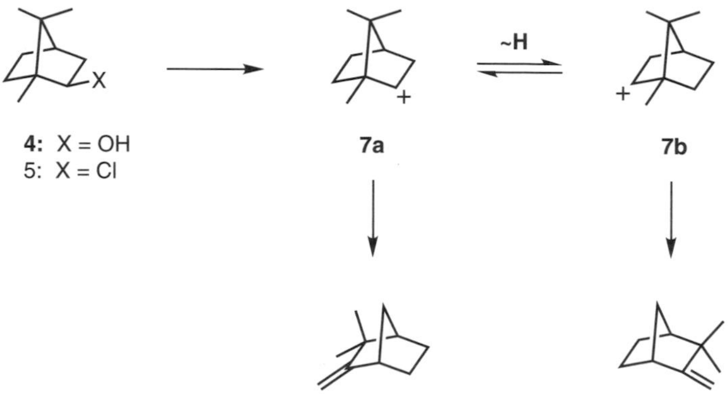

~H

4: X = OH
5: X = Cl

7a **7b**

6a **6b**

Scheme 5.5

5.2.2
A Challenge from Houben and Pfankuch

This theory was not seriously challenged for about six years, until Houben and Pfankuch [8;9] made an unexpected observation during the re-investigation of some earlier work by Passerini on a different topic. To simplify a complicated story, we can concentrate on what this finding was, rather than on the original motivation of Houben and Pfankuch. In this context, their major finding was that the hydrochlorination with rearrangement of camphene-1-carboxamide (**17**, Scheme 5.6) gave 2-chlorocamphane-4-carboxamide **18**, rather than 2-chlorocamphane-2-carboxamide **19**, the product of simple Wagner rearrangement, or 6-chlorocamphane-2-carboxamide **20**, the product of Wagner rearrangement followed by Meerwein-type 6,2-shift. [8;9]

Houben and Pfankuch believed that product **18** was formed by way of a 3,2-methyl shift, sometimes referred to as a "camphene rearrangement of the second kind," an example of which had been discovered in 1927 by Nametkin and Brüssoff. [10] By analogy, this called into question the structure **11** assigned by Meerwein and Wort-

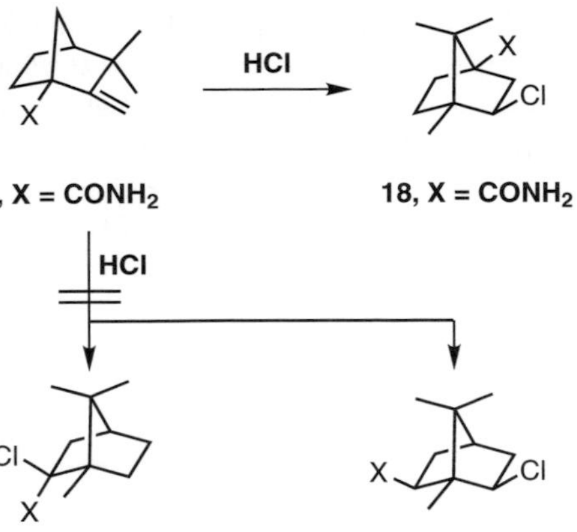

HCl

17, X = CONH$_2$ 18, X = CONH$_2$

HCl

19, X = CONH$_2$ 20, X = CONH$_2$

Scheme 5.6

mann [6] to β-camphordichloride (Scheme 5.3) and the mechanism proposed by Meerwein and Montfort [7] for the racemization of isoborneol **4**, isobornyl chloride **5**, and camphene **6** (Scheme 5.5). Houben and Pfankuch [8;9] and also Bredt [11] recognized that a 3,2 methyl shift in the rearrangement of α–camphordichloride **10** (Scheme 5.7) analogous to that in the rearrangement of camphene-1-carboxamide **17** (Scheme 5.6) would produce not the Meerwein-Wortmann 2,6-dichlorocamphane structure **11** (Scheme 5.3) for β–camphordichloride, but instead the structure 2,4-dichlorocamphane **21** (Scheme 5.7).

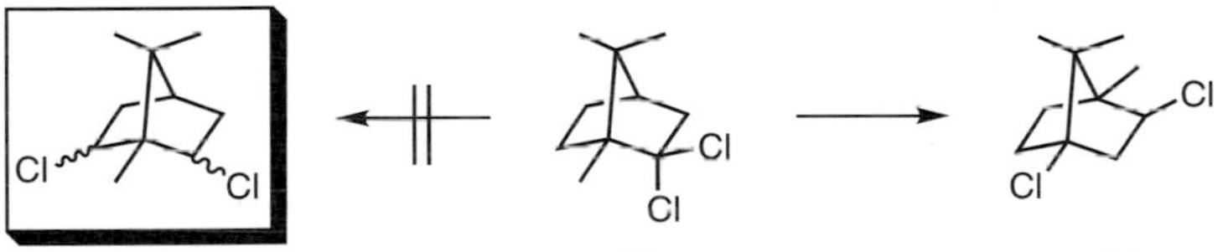

11 "β–camphor-dicloride" 10 α-camphor-dicloride 21 β–camphor-dicloride

Scheme 5.7

Houben and Pfankuch then made the critical mental connection between their result and the isobornyl-camphenehydro set of racemizations, namely that an analogous Nametkin shift (Scheme 5.8) would interconvert the enantiomers of isobornyl chloride **5a** and **5b**.

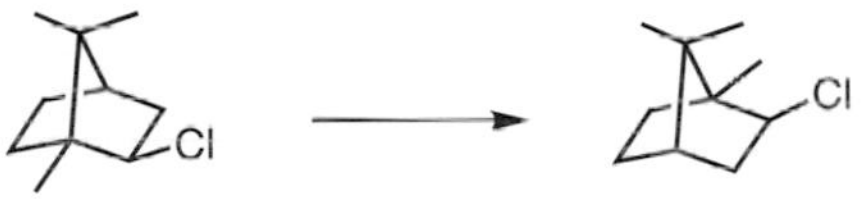

5a 5b

Scheme 5.8

Houben and Pfankuch [12] supported this interpretation with two lines of argument. First, they pointed out a major oversight in the Meerwein-Wortmann discussion of the rearrangements of α-camphordichloride **10** (Scheme 5.3). Meerwein and

Wortmann [6] had prepared this compound from (+)-*1R*-camphor (**1**, Scheme 5.1) and had caused it to rearrange to "α-chlorocamphene" (1-chlorocamphene **12**), which they then reduced to camphene itself **6a**. As we shall see, it is significant that they apparently did not measure the optical rotation of the camphene, but Houben and Pfankuch repeated the experiment and found that the substance was predominantly dextrorotatory. How important this was becomes clear from a comparison of Schemes 5.3 and 5.9.

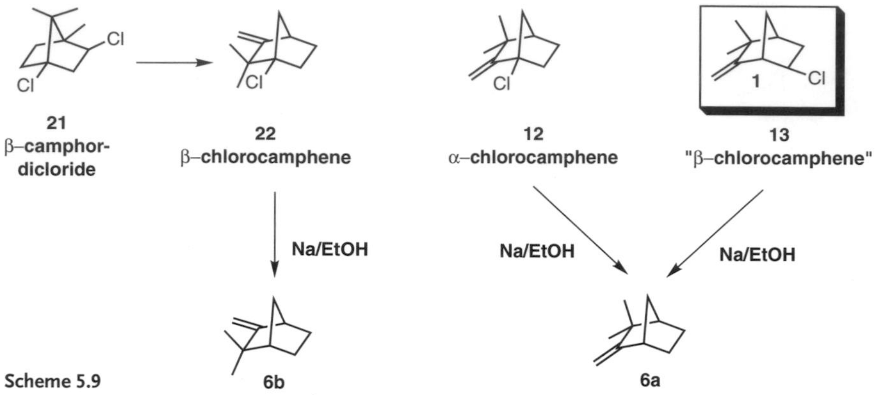

21 β–camphor-dicloride
22 β–chlorocamphene
12 α–chlorocamphene
13 "β–chlorocamphene"
Na/EtOH
Na/EtOH
Na/EtOH
6b
6a

Scheme 5.9

Scheme 5.3 shows that the Meerwein-Wortmann formulation of the rearrangements of α–camphordichloride *1R*-**10** predicts that the camphene derived from *either* α– or β-chlorocamphene (**12** or **13**) would give the *same enantiomer* **6a** of camphene on reduction, whereas the Houben-Pfankuch formulation (Scheme 5.7) of the dichlorocamphane product as **21** and its β–chlorocamphene as **22** predicts (Scheme 5.9) that α–chlorocamphene **12** and β–chlorocamphene **22** should give a *different enantiomers of camphene*: respectively **6a**, which they already had observed to be dextrorotatory (see above), and **6b**, which then would necessarily be levorotatory. Indeed, as they had predicted, Houben and Pfankuch [12]found the camphene from β–chlorocamphene to be levorotatory.

Note that the Meerwein-Wortmann formulation of the β–camphordichloride → β-chlorocamphene rearrangement as **11** → **13** depended upon certain assumptions about the configurations and relative reactivities of the two chlorines of **11**. These assumptions were based on analogy to the assignments at the time to bornyl and isobornyl chlorides which were subsequently shown to be reversed by others in the period 1936–1960. [13] An argument can be made that if Meerwein had known of this reversal, he might have been able to put up a stronger defense of structure **11**, but the point is academic in view of the subsequent findings from Houben's laboratory.

Houben and Pfankuch then did an even more persuasive experiment to show that the structure of "β-camphordichloride" was not 2,6-dichlorocamphane **11**, as Meerwein and Wortmann had proposed, but instead must be 2,4-dichlorocamphane **21** (Scheme 5.10).

Thus Meerwein had found that the action of phenol/potassium acetate on "β-cam-

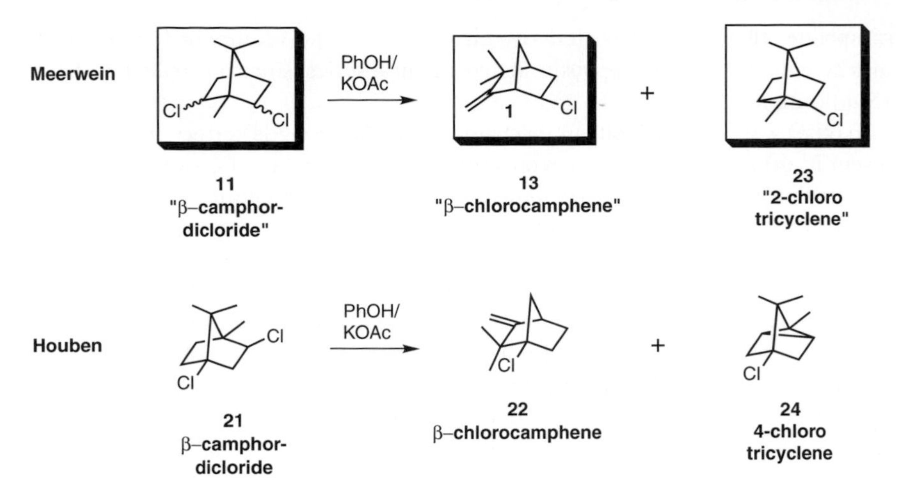

Scheme 5.10

phordichloride" gave, in addition to the olefin β-chlorocamphene, which he had formulated as **13**, another substance which contained no double bonds and which could be reductively dehalogenated to tricyclene. He formulated this chlorotricyclene as 2-chlorotricyclene **23** (Scheme 5.10). This compound is chiral and should have been optically active, but again Meerwein failed to measure the optical rotation. Houben repeated the experiment and showed that the chlorotricyclene was *optically inactive,* and hence the chlorine must lie on the molecular symmetry plane, as in 4-chlorotricyclene **24**. Thus the dichlorocamphane precursor must have one chlorine atom at C_4. This was compelling evidence that Meerwein's formulation of β–camphordichloride as 2,6-dichlorocamphane **11** could not be correct and that the true structure of that substance is 2,4-dichlorocamphane **21**.

Houben and Pfankuch [12] went even further (Scheme 5.11). They recognized that the rearrangement of Scheme 5.7 converted α–camphordichloride (2,2-dichloro-

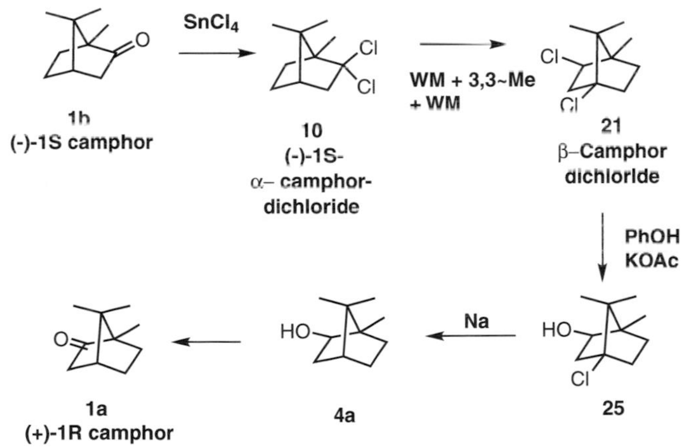

O
SnCl$_4$
Cl
Cl
WM + 3,3~Me
+ WM
Cl
Cl
1b
(-)-1S camphor
10
(-)-1S-
α– camphor-
dichloride
21
β–Camphor
dichloride
PhOH
KOAc
O
HO
Na
HO
Cl
1a
(+)-1R camphor
4a
25

Scheme 5.11

camphane) **10**, which had been derived from *D* (*i.e.* 1*R*)-(+)-camphor **1** (Scheme 1), into **21**, a derivative of the opposite stereochemical series, since it is related to *L* (*i.e.* 1*S*)-(-)-camphor.

In other words, if the Houben mechanism of Scheme 5.7 is correct, it would represent in principle a way of interconverting the enantiomers of camphor. Houben and Pfankuch [12] immediately demonstrated this experimentally in spectacular fashion (Scheme 5.11). Thus 1*S*-(-)-camphor **1b** was converted to α–camphordichloride **10** of the corresponding stereochemical series, which upon being subjected to the rearrangement of Scheme 5.7, provided 2,4-dichlorocamphane *1R*-**21**. Hydrolysis of the secondary chlorine gave 4-chloroisoborneol **25**, which upon reductive dehalogenation followed by oxidation of the intermediate 1*R*-isoborneol **4a**, gave 1*R*-(+)-camphor **1a**. The overall result was the same as if a reaction conceivable only in fantasy had occurred and the camphor bridge had been pushed right through the molecule in response to an irresistible thrust along the direction of the dark arrow (Scheme 5.12), thereby inverting the stereocenters at both bridgeheads.

Scheme 5.12

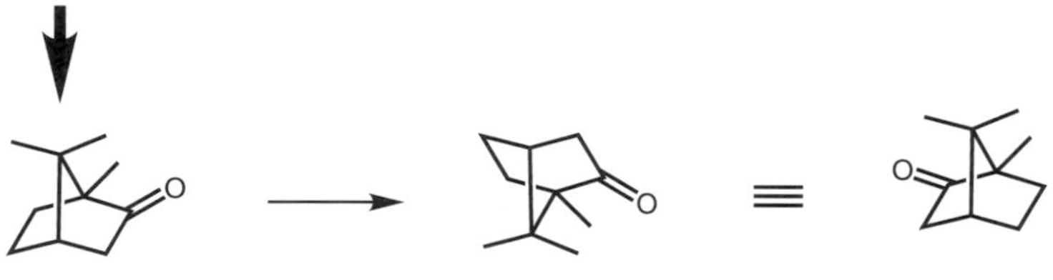

1b
(-)-1S camphor

1a
(+)-1R camphor

This transformation absolutely stunned the chemists of that day. One can still sense the astonishment and admiration in the minds of Simonsen and Owen many years later in their 1949 book [14] on the chemistry of the terpenes.

5.2.3
Death and Resurrection of Theories. The Non-Vicinal Shift Theory Comes Back

One might suppose that these developments were sufficient to refute the Meerwein 2,6-shift theory once and for all. Houben and Pfankuch summarized their findings on the mechanism of racemization in the camphene series in the words "*This is the elucidation of the observed racemization,* not the Meerwein theory of a 2,6-oscillation of a halogen atom." But is would be wrong to conclude that the theory therefore had no validity. One must remember the case pointed out by Meerwein, the rearrangement of fenchol to isofenchol (Scheme 5.4). This reaction, which Meerwein suggested also to be an example of the 2,6-shift mechanism at work, cannot easily be explained by a Houben-style Nametkin *cum* Wagner mechanism. Moreover, as has been detailed elsewhere, [15] twenty years later others were able to show by isotopic tracer methods that *both* mechanisms were active in the racemization of camphene and that 2,6-shift also played a prominent role in other reactions. [16–19] Houben thus interpreted his findings in the case of the dichlorocamphanes too broadly. What he did-n't realize was that although the 2,6-shift could not account for the structure of the rearrangement product of α–dichlorocamphane, one could not rule out the possibility of 2,6-shift in other reactions, specifically those in the camphene-isobornyl systems. In other words, refutation of supposed experimental evidence for a theory does not conclusively refute the theory itself.

This case shows how elusive a goal the true refutation of a theory can be. In Popper's philosophy, refutation is the driving force of scientific advance, because confirmations of a theory logically cannot guarantee that no refutation eventually will be found. *But we now have just illustrated the inverse situation, that is, apparent refutations of a theory cannot logically guarantee that no confirmation eventually will be found.* Houben's insightful analysis of the racemization phenomena in the camphene series, and his masterly handling of the stereochemical issues involved in working out the consequences of the analysis, had produced as formidable an example of classical Popperian refutation as could be imagined. Meerwein had no response, and for decades, the chemical community seemed to be convinced that there were no 2,6-shifts and that this was the end of the theory. Yet the theory refused to die and ultimately was shown to be valid in other cases.

5.2.4
Further Resurrections

Two similar resurrections are found in other examples I have described elsewhere: [20] One is the survival of the Woodward-Hoffmann orbital symmetry rules despite Oosterhoff's demonstration that the authors' extension of them to photochemical reactions improperly assumed that the reactive state was the first excited state. [21] Another is the demonstration by Futaki that the Dewar π–complex mechanism for the dienone-phenol rearrangement, which had been ruled out in Woodward's ingeniously designed test case, nevertheless applied to other cases.

A similar case is found in an article by Martin Gardner [22] entitled "A Skeptical Look at Karl Popper." This treatment reaches conclusions similar to those of the present chapter. I quote one telling passage.

> "Falsifications can be as fuzzy and elusive as confirmations. Einstein's first cosmological model was a universe as static and unchanging as Aristotle's. Unfortunately, the gravity of suns would make such a universe unstable. It would collapse. To prevent this, Einstein, out of thin air, proposed the bold conjecture that the universe, on its pre-atomic level, harbored a mysterious, undetected repulsive force he called the 'cosmological constant.' When it was discovered that the universe is expanding, Einstein considered his conjecture falsified. Indeed, he called it 'the greatest blunder of my life.' Today, his conjecture is back in favor as a way of explaining why the universe seems to be expanding faster than it should. Astronomers are not trying to falsify it; they are looking for confirmations."

We see here examples related to one of Feyerabend's major themes: [23]

> "There is no idea, however ancient and absurd, that is not capable of improving our knowledge. The whole history of thought is absorbed into science and is used for improving every single theory."

This perhaps overstates the case, but the idea that we should not be too quick to discard existing theories is undeniable.

These examples reveal a common difficulty in the falsificationist program. We are advised to design tests of theories that will expose them to the most stringent refutationally motivated experiments. But a theory makes *discrete* predictions: If you do so-and-so under these conditions, then such and such will result. A "crucial" test experiment thus is a singular, unique event. It cannot address all of the possible predictions of the theory. Thus, the demonstration that a particular prediction of the theory fails under a specific set of circumstances cannot be taken to invalidate the theory's applicability to other circumstances. The temptation to over-generalize was, as we have seen, the source of Houben's mistaken conclusion that there was no 6,2-shift.

5.3
A Premature Death Certificate for Willstätter's "Small Molecule" Theory of Enzymes?

Ideally, if we accept an experimental refutation as valid, we tacitly assume that the criteria of refutation we applied were valid. Sometimes, however, later knowledge or even just further reflection may reveal that this assumption was dubious. In other words, the strength of the refutation may depend on the particular historical period in which it occurred.

It is in this context that we now examine a case of a quickly accepted refutation based on grounds that were *ambiguous at the time*. Only much later did new evidence of a completely different kind dispel the uncertainty. This example comes from research during the last century on the nature of enzymes.

The enzymes are ubiquitous organic catalysts which modulate virtually all chemical reactions in living organisms. Even reactions whose spontaneous, uncatalyzed *in vitro* reaction is fast, such as the conversion of carbonic acid to carbon dioxide and water, is controlled in the biological cell by an enzyme. In fact, Richard Willstätter maintained [24] that "we may regard life as a system of cooperating enzymatic reactions."

To the modern chemist, the idea that enzymes are proteins is so commonplace that we give no thought to alternatives. We are aware that our colleagues in molecular biology and biochemistry conduct their entire enterprise on this conviction. Even a chemist whose work lies in other areas would dismiss as prattle any suggestion to the contrary. Imagine the reaction today to anyone foolish enough to claim that although it does seem to be the case that enzymes are *closely associated* with proteins, there is no definitive proof that they *are* proteins. After all, even if we are not fully cognizant of the mass of evidence from biophysical chemistry and genetics, a moment's reflection will remind us that not only can one determine the structures of enzymes by nuclear magnetic resonance (NMR) spectroscopy or X-ray crystallography, one actually can observe the structure of enzyme-substrate complexes. This allows crystallographers to take a still photograph of the partners collaborating in the very act of enzyme catalysis. Moreover, using time-resolved X-ray crystallography, a technique that until recently was only a fantasy, one can follow this process in real time, producing, as it were, a motion picture of the event.

What many chemists nowadays may not realize is how controversial the chemical nature of enzymes was for a substantial time, roughly the decade 1920–1930, when two major theories were in contention. A historical analysis of how this issue was fought out, and how an overwhelming consensus favoring one of them was reached, can be instructive with respect to the methods and meaning of refutations in science. As we shall see, although the choice of the majority eventually was compellingly supported by powerful evidence, it was accepted well before that, at a time when escape routes for the small molecule theory still were open.

During the first third of the 20th century, biochemists carried out their research on enzymes on solutions of these materials obtained from biological sources. Initially, these solutions were of uncertain concentration and homogeneity. One can only marvel at the succession of discoveries made with relatively crude enzyme preparations. It was a period in which the mechanistic pathways of metabolic processes began to be traced out. Biological events such as fermentation, digestion, growth, and the synthesis of bodily constituents in plants and animals became clearer. As a consequence, much knowledge was gained of the treatment and prevention of disease. [25]

Eventually, methods for the analysis of enzyme preparations and for concentration and eventual purification were developed. But in the period 1920–1930, immediately preceding those advances, the nature of enzymes themselves was problematic.

5.3.1
What Are Enzymes? Willstätter's Small Molecule Theory

Two major theories were in contention: one that enzymes are proteins, which is the surviving dogma, and the other that an enzyme is made up of a small molecule, which embodies the catalytic activity, and a protein which serves as the carrier of the active site and accounts for the specificity. [25;26] The latter theory we now believe to be inadmissible. The major proponents of the small molecule theory were Richard Willstätter [24] and members of his school, [27] most of them his former students, whose ideas grew out of their work on yeast invertase and several other enzymes.

Like known non-enzymatic proteins, the enzyme preparations gave positive chemical color tests for various protein amino acids, especially tryptophan, the response becoming stronger as the enzymatic activity increased through cycles of the purification methods of the time. However, Willstätter apparently convinced himself that the reactive moiety of the enzyme was a natural, non-proteinaceous small molecule. The basis for this was a series of experiments [28] on enzymes prepared by precipitation from solution and prolonged dialysis. An important example was the proteolytic enzyme pepsin, prepared from an extract of calf stomach mucosa. This experiment was essentially a repetition of a much earlier one by Sundberg, [29] in which that author reported that although his preparations were protein-free, they were still catalytically active. Similarly, Willstätter and Rohdewald [28] showed that although their own preparations were apparently "protein-free" as judged by negative color tests, *some catalytic activity nevertheless remained.*

Others have described [30–33] Willstätter's work during this period as carried out under most unconventional, and probably less than ideal, circumstances. In 1924, sensitive (some [31–33] say perhaps over-sensitive) to what he felt were strong currents of anti-Semitism among the faculty and students, he had resigned his professorship at the University of Munich. He had given up all his laboratory space except for one small area made available by his successor, Heinrich Wieland, and occupied by his only remaining co-worker, Margarete Rohdewald. He had vowed never to set foot on campus again, and he communicated with her by telephone.

Willstätter's position on the enzyme issue was difficult to attack. He maintained that whenever enzyme preparations gave positive color tests for protein, they were due to the proteinaceous carrier adhering to the supposed small molecules responsible for the actual enzymatic catalysis. Presumably, but not explicitly, he seemed to think that in the so-called "protein-free" enzyme solutions, all the protein had been removed, and the catalytic activity remaining could be ascribed to the small molecule moiety. This showed, he claimed, that the active part of the enzyme was not a protein.

At the same time, he was never able to isolate or identify just which small molecule was at work. Nevertheless, he stubbornly held to this position despite many critical objections.

Of course, as was later pointed out by his opponents, what his argument lacked was a quantitative measure of the sensitivity of the tests. If the enzyme was sufficiently active even at concentrations below the limit of detectability of the protein tests, a proteinaceous enzyme still could be present in the "purified" preparation. One such objection was raised by Henry Tauber [38] at the New York Homeopathic Medical College and Flower Hospital. He prepared solutions of highly purified rennin and purified, actually crystalline, pepsin and was able by simple dilution of these to obtain solutions for which color tests for protein were negative, but the catalytic activity was still readily detectable. He concluded that the color tests were simply not sensitive enough to detect small amounts of protein. According to this interpretation, the problem seemed to be that Willstätter's tests simply were of insufficient dynamic range.

5.3.2
Crystallization of Enzymes. Enzymes Are "Pure" Proteins

Perhaps the most influential studies of that period pointing to the protein nature of enzymes were initiated by the successful crystallizations of enzymes. [34;35] The first came in 1926, [36] when James B. Sumner of Cornell University (Figure 5.1), working with extracts from jack bean meal, crystallized urease, an enzyme that catalyzes the *in vivo* hydrolysis of urea. In 1930, John H. Northrop of the Rockefeller Institute

Figure 5.1 James B. Sumner, Cornell University. © The Nobel Foundation.

Figure 5.2 John H. Northrop, Rockefeller Institute. © The Nobel Foundation.

(Figure 5.2) succeeded in crystalling pepsin, a proteolytic enzyme isolated from swine stomach tissue. [37]

Several kinds of experimental evidence supported the conclusion that these and subsequently obtained crystalline enzymes were proteins. One was the observation that they, like other proteins, were hydrolyzed in the presence of proteolytic enzymes. In fact, some isolated enzymes in solution were thought to be capable of catalyzing their own hydrolysis. Further, the observed molecular weights had values from tens to hundreds of thousands, like many other proteins. Solutions of crystalline enzymes gave upon hydrolysis typical mixtures of the known amino acids, such as glycine, glutamic acid, tyrosine, tryptophan, etc.

Despite Willstätter's gallant rear-guard action, both Sumner and Northrop were quickly convinced that the enzymes were proteins and nothing else. Tauber [38] quotes from a private letter from Northrop to Willstätter which is quoted in the Willstätter-Rohdewald paper of the same year [28] and which leaves no doubt of Northrop's conviction that "the proteolytic properties and the protein properties are both attributes of the same molecule."

We must keep in mind a warning articulated by Fruton and Simmonds: [39]

"It would be incorrect, however, to conclude, from the fact that a protein has been crystallized, that it has been obtained as a single pure substance in the same sense that one is accustomed to think of a sample of crystalline benzoic acid. It will be evident from the subsequent discussion of the chemistry of proteins that they are extremely complex in structure, and the definition of their purity is an experimental problem of the greatest difficulty."

One does not have to speculate unduly to make the plausible conjecture that much of Northrop's conviction of the protein nature of enzymes came from the beautiful work which he and Moses Kunitz carried out on the phase rule test for homogeneity of proteins. [40–42] The principle of the method is introduced by Kunitz and Northrop:

> "The demonstration of purity in the case of proteins by analytical methods is therefore extremely difficult. Fortunately a definite criterion, experimentally and theoretically , is offered by Willard Gibbs' [s] phase rule. This theory furnishes the explanation for the familiar fact that constant melting point is a very sensitive criterion of purity. Proteins, however, do not melt without decomposition, but solubility determinations furnish an analogous test which may be applied to proteins."

Sorensen seems to have been the first to apply the method to the determination of protein homogeneity. [43] The method makes use of the postulate of phase rule theory that in a system at equilibrium at constant temperature and pressure, the composition of the phases is fixed and cannot be varied by changing the quantity of the phases present. This has the consequence that at constant temperature and pressure, the solubility of a solid pure substance in a specified solvent will increase until it reaches its solubility limit, where any excess added will not dissolve further, no matter how much is added. Up to the solubility limit, the slope of the plot should be 1.00, since in that region, all of the added solid dissolves. Beyond that limit, adding excess solid will not increase the amount of substance in solution. In the ideal case, this

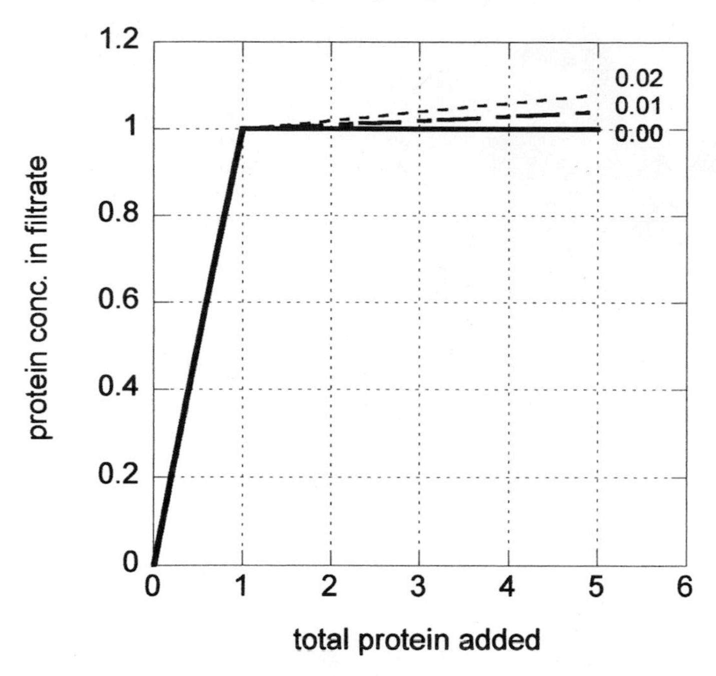

Figure 5.3 Schematic representation of the solubility of a protein with no impuritiy, with 1% impurity, and with 2% impurity. The slope of the line to the right of the break point gives the fraction of impurity.

phase transition, from homogeneous solution to solution containing undissoved solid, is sharply defined. Figure 5.3 [41] which exemplifies the behavior of a "pure" sample, shows the solubility of a crystalline protein in an aqueous salt solution as a function of the total amount of solid added. Note the sharp intersection point, beyond which the plot is level.

Figure 5.3 also shows schematically the behavior of an almost but not quite pure sample comparable to that observed for γ-chymotrypsin. [41] Note that the slope of the line beyond the break point is not quite zero, and in fact, the value of the slope is 0.02 (2%), corresponding to the fraction (or percentage) of impurity. In this case, the analysis is quite sensitive, and even as little as 1% of impurity could have been detected. In cases with larger amounts of impurity, the break is less pronounced, and many points must be taken in the region of the break in order to define the slope accurately. The analysis can be extended to three-component mixtures, but this will not concern us here.

In most cases, a 0.1% or smaller contaminant would be difficult to detect by the phase rule method, which was the most sensitive one then available. For the purpose of the discussion to follow, we may take that as the point at which our ability to be sure that no contaminant is present would fade out.

Taken together with results from other methods of analysis, such as electrophoresis and ultracentrifugation, the phase rule studies had a great impact on the thinking

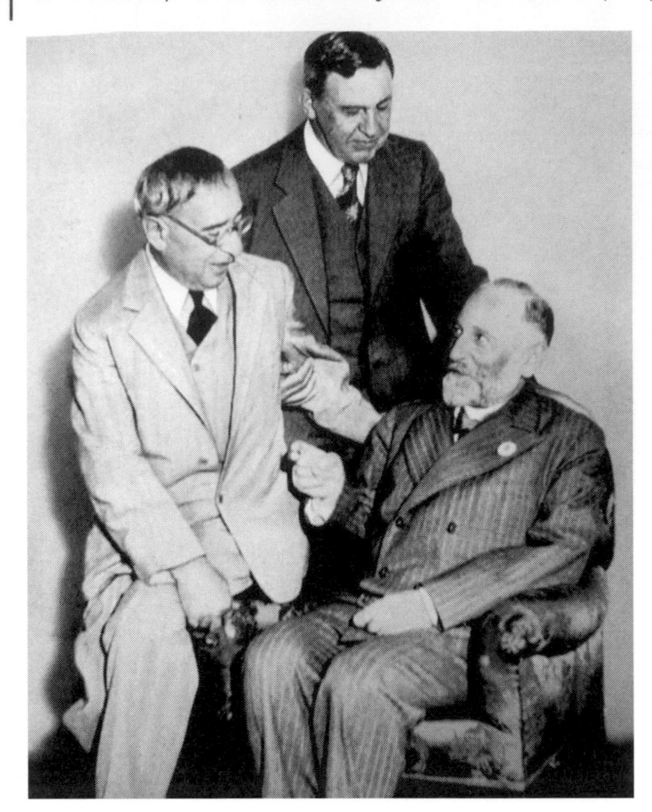

Figure 5.4 Three attendees at the 1933 Chicago meeting of the American Chemical Society: (left to right) Moses Gomberg, Vladimir Ipatieff, and Richard Willstatter. Reproduced from *Aus Meinem Leben*, by R. Willstätter, 1949, Verlag Chemie, with permission of Verlag Chemie and Wiley-VCH.

of the biochemical community. The perception became widespread that Willstätter's proposals could not be correct. Crystalline enzymes were "pure" compounds by all the criteria used, and in a short time, few chemists were left who supported the idea that catalytic activity in Willstätter's so-called "protein-free" enzymes could be due to a small amount of a small-molecule active entity as a contaminant in the preparations.

5.3.3
An Incomplete Refutation. How Pure Are Proteins?

It is true that by this time, the Willstätter faction was fast being crowded into an ever-narrowing space for maneuver. In fact, it was not long before their numbers dwindled almost to extinction. Resistance, even from Willstätter himself, seemed to be wavering. In his Willard Gibbs Medal address at the meeting of the American Chemical Society (see Figure 5.4) in September of 1933, [30] he said:

> "Whether we deal with such tentative explanations, or with the controversial protein nature of enzymes, I feel that it is not important for the scientist whether his own theory proves the right one in the end. Our experiments are not carried out to decide whether we are right, but to gain new knowledge. It is for knowl-

> edge's sake that we plow and sow. It is not inglorious at all to have erred in theories and hypotheses. Our hypotheses are intended for the present rather than for the future. They are indispensable for us in the explanation of the secured facts, to enliven and to mobilize them and above all, to blaze a trail into unknown regions towards new discoveries."

Later, he quotes J.J. Thomson: "A scientific theory is a tool and not a creed."

Yet in retrospect we can see that the switch in the community's attitude, from skepticism of the small molecule theory to complete rejection of it by an overwhelming majority, cannot be said to have resulted from some kind of logical or unequivocal or final refutation. A few simple calculations will show that the experimental evidence in the 1930s and 1940s still left a good deal of room for small-molecule enzyme die-hards to salvage some credibility for their cause, had they chosen to do so. The argument would go as follows:

Obviously, if the enzyme sample is a mixture consisting preponderantly of catalytically inactive protein and a small amount of an active small molecule, the actual catalytic activity per microgram of the pure small molecule itself must be much greater than that of the same weight of the mixture, where the active catalyst is diluted. Thus, if one could isolate a pure sample of the actual small molecule catalyst, its characteristic catalytic rate would have to be large enough to produce the catalytic effect observed with the mixture. However, one cannot simply postulate ever-increasing activity for the alleged small molecule catalyst, because no thermal rate constant can exceed the rate of a molecular vibration, which is about 10^{12-13} s^{-1}. Thus, the advocates of a small molecule origin of enzyme catalysis do not have an indefinitely expanding field of conjecture in which to rescue their theory. The key question is how active does the small molecule catalyst have to be to match the observed rate? If it can be shown to be less than the theoretical limit, the theory is at least *logically* defensible in the sense that it does not directly conflict with experimental fact.

It is easy to show that a small molecule catalyst present to the extent of 0.1% as a contaminant in a sample of catalytically inactive protein would not have to exceed this limit in order to produce typical catalytic effects. We make the usual assumption that the kinetics of the transformation of a substrate S to a product P mediated by an enzyme E follow the Michaelis-Menten kinetics [44;45] (see Scheme 5.13).

$$\mathrm{E + S} \underset{k_2}{\overset{k_1}{\rightleftharpoons}} \mathrm{ES} \xrightarrow{k_3} \mathrm{E + P}$$

Scheme 5.13

The activity of an enzyme may be characterized as the so-called "turnover number," that is, the number of substrate molecules per unit time converted into product per molecule of enzyme. This is given kinetically by the unimolecular rate constant k_3 for the decomposition of the enzyme-substrate complex. This in turn can be derived experimentally from saturation experiments, as shown in the in the references [44;45].

Typically, the observed rate, $-d\,[S]/dt$, of an enzyme-mediated reaction increases with increasing substrate concentration until the enzyme is "saturated". Further in-

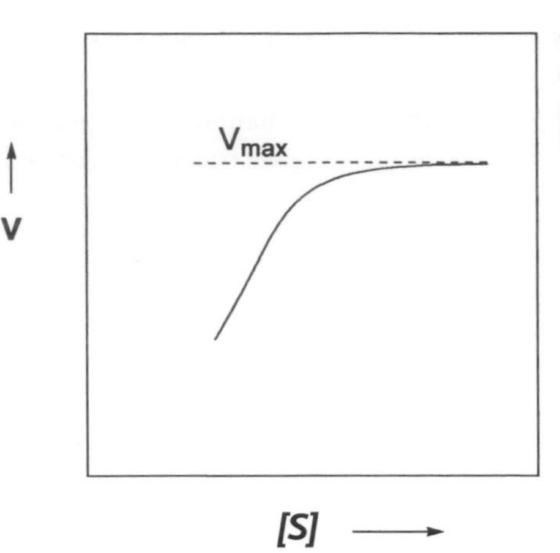

Figure 5.5 Schematic diagram for a typical enzyme reaction. The instantaneous rate is given by *V* and the substrate concentration by S.

crease of the substrate concentration has no further effect on the rate, so that a plot of the observed rate as a function of *[S]* rises at first and then levels off to a steady value (see Figure 5.5).

This rate is called V_{max}; it is related to k_3 by the equation $V_{max}/[E_t] = k_3$, where $[E_t]$ is the total concentration of enzyme. Turnover numbers for known enzymes range in value from about 1 s^{-1} or less to about 600,000 s^{-1}.

Let's imagine how we would determine the turnover number for the important proteolytic enzyme chymotrypsin. We would want to choose conditions convenient for ourselves, so that the half-life of substrate under our conditions would be of the order of hours, a rate neither intimidatingly fast nor tediously slow to measure. For the case at hand, these conditions would be met conveniently from an observed rate V_{max}, for example, about 10^{-4} Ms^{-1}, so that about 100 μmol of substrate per second was being converted to product. Suppose we further assume that the enzyme concentration E_t is 10^{-6} *M*, a typical value for many of such experiments. The apparent turnover number k_3 then would be $[10^{-4}\ Ms^{-1}]/[10^{-6}\ M]$ or 10^2 s^{-1}, which is the reported value for chymotrypsin.

Suppose we now consider the possibility that the activity of chymotrypsin is caused by a small molecule contaminant. Could a conveniently measurable rate be observed if, for example, the contaminant was present to the indetectably small extent of 0.1 % in the enzyme sample? Since $V_{max}/[E_t] = k_3$ (see above), if E_t is actually only 0.001 of the nominal enzyme concentration, the true $[E_t]$ is no longer 10^{-6} *M* but rather 10^{-9} *M*. The true k_3 is no longer 10^2 s^{-1} but instead is 10^5 s^{-1}, which is still far below the theoretical rate limit of 10^{12-13} s^{-1} for a unimolecular reaction (see Figure 5.6).

Generally, the ratio of the "true" turnover number for a contaminating catalytic species to the apparent turnover number is just the reciprocal of the fractional contamination. For example, if we take 0.001 (0.1%) as the fractional contamination, the "true" turnover number would have to be 1000 times that of the apparent one. Thus, even in the case of one of the most active enzymes, carbonic anhydrase (apparent turnover number 6 x 10^5 s^{-1}), the "true" turnover number of 6 x 10^8 s^{-1} for the putative 0.1% contaminant still would fall short of a violation.

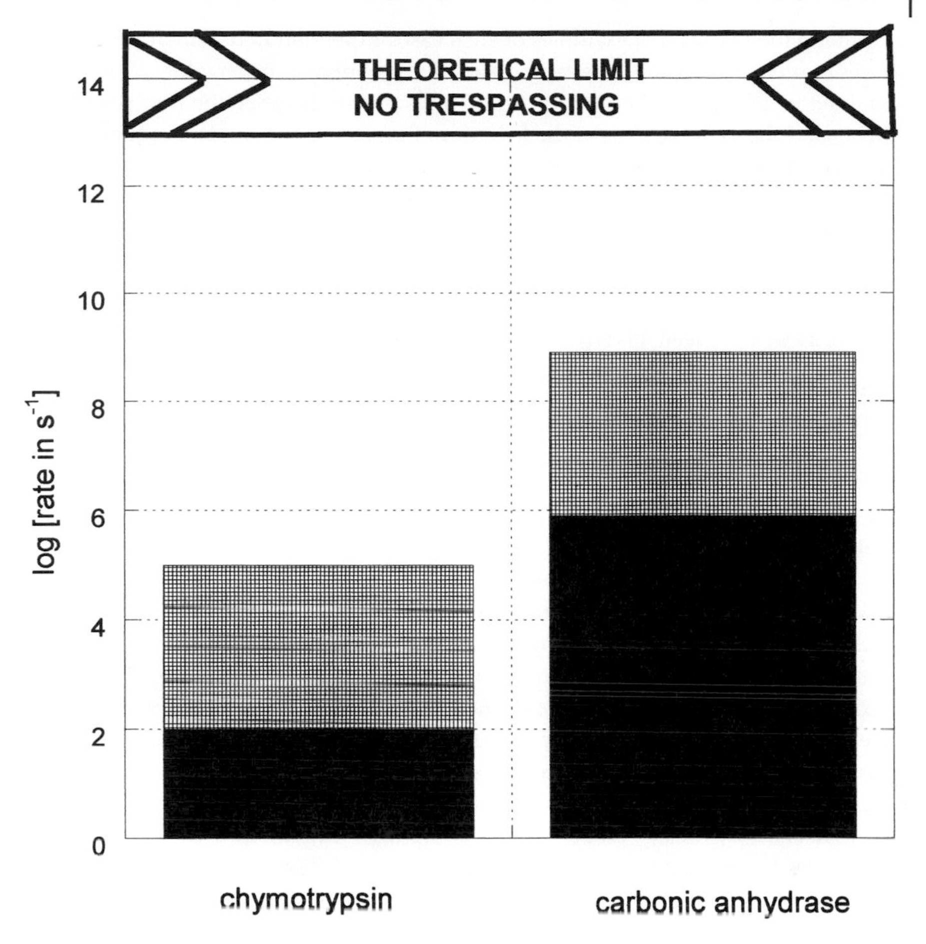

Figure 5.6 Log [turnover number] for two enzymes. The black regions represent the observed values based upon the assumption that the enzyme is pure. The checkerboard regions are the increments in log [turnover number] needed if the active enzyme is present as an impurity to the extent of 0.1 %. Note that even the enhanced rates still fall short of the theoretical limit.

What these simple calculations show is that neither the crystallization of proteins nor the claim of their "purity" sufficed *logically* to refute Willstätter's small-molecule enzyme theory. Nevertheless, as we have seen, the remaining adherents of the theory in the biochemical community for the most part quickly abandoned it, and Willstätter and his supporters did not fight their battle to the bitter end. This kind of outcome is by no means uncommon in science. From the standpoint of an observer monitoring the flow of scientific history, it has the *effect* of a refutation: workers in the field recognize that the old theory is no longer capable (if it ever was) of leading to advances, and the mainstream of science branches off into new channels.

Among the many other examples of such cases I mention here Williamson's synthesis of unsymmetrical ethers, which was an experimental cornerstone of the structural theory of organic chemistry. It tested a distinction between the new structural theory and the existing ideas of the behavior of organic compounds. From our later vantage point, we can find grounds for scepticism that the distinction was really airtight, but those reservations were either not recognized at the time or were simply disregarded. As Rocke puts it [46]

> "Whether or not retrospective logical analysis can sustain Williamson's claim to have provided an irrefutable test, empirical history supersedes logical necessity, for as we have seen, his argument was soon universally regarded as compelling."

It is my impression that the history of science contains as many such virtual refutations as logically impeccable ones, if not more.

5.4 Conclusion

We cannot assume that there is such a thing as a *permanent* refutation, one that is destined to last forever. At most, we can say that we have an apparent, or effective, or virtual refutation. First, any refutation is subject to the *ceteris paribus* reservation (see Chapter 6) and may have to be abandoned if later events show it to have been violated. Other refutations ultimately themselves may be overthrown by later demonstrations that they are based on erroneous thinking or faulty experimental execution.

One may say that refutation, like discovery itself, often is not a temporally well-defined, decisive, "crucial", event. It often is a *process*, extending over years or even decades. Whatever their attitude toward the doctrine of falsificationism, in practice, scientists often violate its precepts without formal justification. Despite the existence of a supposed falsification of a theory, scientists obviously use other criteria to decide whether to continue using it. The decision often is based on their evaluations of its continuing heuristic power, not on its *logical* defensibility.

5.5 Acknowledgment

Sections 5.2–5.2.4 of this chapter are expanded from discussions I have given elsewhere: J.A. Berson, in *Chemical Creativity*, Wiley-VCH, Weinheim, 1999, pp. 132ff and *Molecular Rearrangements*, P. de Mayo, ed. 1963, Interscience, New York, Vol. I, pp. 155–160.

5.6
References and Notes

1 K. R. Popper, *The Logic of Scientific Discovery* K. R. Popper, J. Freed, L. Freed, translators, Routledge, London **1959**. p. 277. From *Logik der Forschung*, the original German edition of 1934 (reprinted 1992). Translators K.R. Popper, J. Freed, L. Freed.

2 T. Kuhn, *The Structure of Scientific Revolutions, 2nd ed., enlarged,* University of Chicago Press, Chicago, IL, **1970**.

3 P. Galison, *How Experiments End,* University of Chicago Press, Chicago, IL, **1987**. Chapter 1

4 H. Meerwein, K. van Emster, *Ber.* **1920**, *53,* 1815.

5 H. Meerwein, K. van Emster, *Ber.* **1922**, *55,* 2500.

6 H. Meerwein, R. Wortmann, *Ann* **1924**, *435,* 190. For reasons that would be considered incorrect today, the authors assigned the 2-endo-6-exo configuration to the supposed product **11**. The point is moot, since the structure **11** itself is incorrect.

7 H. Meerwein, F. Montfort, *Ann.* **1924**, *435,* 207.

8 J. Houben, E. Pfankuch, *Ann.* **1930**, *483,* 273.

9 J. Houben, E. Pfankuch, *Ann.* **1931**, *489,* 193.

10 S. Nametkin, L. Brüssoff, *Ann.* **1927**, *459,* 144.

11 J. Bredt, *J. Prakt. Chem.* **1931**, *[2]131,* 144.

12 J. Houben, E. Pfankuch, *Ann.* **1933**, *501,* 219.

13 J. A. Berson, *Molecular Rearrangements* (Ed.: P. de Mayo), Wiley-Interscience, New York, N.Y **1963**. Part I, p. 123ff. A review.

14 J. Simonsen, L. N. Owen, *The Terpenes* Cambridge University Press, Cambridge, UK **1949**.v. II p. 522.

15 J. A. Berson, *Chemical Creativity: Ideas from the Work of Woodward, Hückel, Meerwein, and Others* Wiley-VCH, Weinheim, Germany **1999**. p. 135.

16 W. v. E. Doering, A. P. Wolf, *Perfumery Essent. Oil Record* **1951**, *42,* 414.

17 J. D. Roberts, C. C. Lee, *J. Am. Chem. Soc.* **1951**, *73,* 5009.

18 J. D. Roberts, J. A. Yancey, *J. Am. Chem. Soc.* **1953**, *75,* 3165.

19 W. R. Vaughan, R. Perry, Jr., *J. Am. Chem. Soc.* **1953**, *75,* 3168.

20 J. A. Berson, *Chemical Creativity: Ideas from the Work of Woodward, Hückel, Meerwein, and Others,* Wiley-VCH, Weinheim, Germany, **1999**.

21 *see however* J. Michl, *J. Am. Chem. Soc.* **1971**, *93,* 523.

22 M. Gardner, *Skeptical Inquirer* **2001**, *July/August,* 13.

23 P. Feyerabend, *Against Method* New Left Books, New York **1975**. p. 5. Reprinted in 1997 by Verso from the third edition of 1993.

24 R. Willstätter, *Problems and Methods in Enzyme Research,* Cornell University Press, Ithaca, NY, **1927**. Based upon the George Fisher Baker Non-Resident Lectures at Cornell University, 1927.

25 J. S. Fruton, *Proteins, Enzymes, Genes: The Interplay of Chemistry and Biology,* Yale University Press, New Haven, CT, **1999**.

26 C. Tanford, J. Reynolds, *Nature's Robots: A History of Proteins,* Oxford University Press, New York, NY, **2001**. Chapter 15.

27 E. Waldschmidt-Leitz, *Ann. Rev. Biochem.* **1932**. v. I, p. 69.

28 R. Willstätter, M. Rohdewald, *Z. physiol. Chem.* **1932**, *108,* 258.

29 C. Sundberg, *Z. physiol. Chem.* **1885**, *9,* 319.

30 R. Willstätter, *Aus Meinem Leben* Verlag Chemie, Weinheim **1949**. p. 355.

31 B. Witkop, *Med. Res. Revs.* **1992**, *12,* 195.

32 F. Litten, *Der Rücktritt Richard Willstätters 1924/25 und seine Hintergründe: Ein Münchener Universitätsskandal?,* Institut für Geschichte der Naturwissenschaften München, Munich, **1999**.

33 R. Huisgen, *Nachricht. Chem.* **2001**, *49,* 30.

34 J. H. Northrop, R. M. Herriott, *Ann. Rev. Biochem.* **1938**, *VII,* 37. A review.

35 J. H. Northrop, M. Kunitz, R. M. Herriott, *Crystalline Enzymes,* Columbia University Press, New York, **1948**. A review.

36 J. B. Sumner, *J. Biol. Chem.* **1926**, *69,* 435.

37 J. H. Northrop, *J. Gen. Physiol.* **1930**, *13,* 739.

38 H. Tauber, *J. Biol. Chem.* **1932**, *99*, 257.

39 J. S. Fruton, S. Simmonds, *General Biochemistry* John Wiley, New York **1958**. p. 27, 207.

40 J. H. Northrop, M. Kunitz, *J. Gen. Physiol.* **1930**, *13*, 781.

41 M. Kunitz, J. H. Northrop, *Cold Spring Harbor Symp. Quant. Biol.* **1938**, *6*, 325.

42 T. Shedlovsky, *Ann. N.Y. Acad. Sci.* **1943**, *48*. A review.

43 S. P. L. Sorensen, *Compt.rend. trav. Lab. Carlsberg* **1925**, *15*, 1.

44 J. S. Fruton, S. Simmonds, *General Biochemistry* John Wiley, New York **1958**. p. 235ff.

45 L. Stryer, *Biochemistry* W. H. Freeman, San Francisco **1974**. p. 124ff.

46 A. J. Rocke, *The Quiet Revolution: Herman Kolbe and the Science of Organic Chemistry*, University of California Press, Berkeley, CA, **1993**. California Studies in the History of Science, J.L. Heilbron, series editor: p. 246.

6
Farewell to Stasis, Welcome to Metamorphosis: Refutation of Kekulé's Rule of Minimal Structural Change in the Discovery of Molecular Rearrangements

"Chemistry is thus the study of the material transformations of matter. Its essential object is not the substance which exists, but rather its past and its future. The relationship of a substance to that which it was and to that which it can become form the characteristic purpose of chemistry."
August Kekulé, Lehrbuch der Organischen Chemie, [1;2] 1861, Vol. 1, p. 1.

6.1
Kekulé's Rules

As most chemists know, Kekulé's seminal paper of 1858 [3] describes how the structures of organic molecules can be built up conceptually by connection of carbons using one of the four valences of each. The remaining three valences of each are fulfilled in making bonds to further carbons or to other atoms. (Note that in this paper, Kekulé did not treat the question of multiple carbon-carbon bonds). However, the paper also contains two further rules which amount to guidelines for thinking about chemical transformations ("Metamorphosen" in Kekulé's word).

The first describes a simple change (*"einfache Metamorphose"*), which today might be called a functional group change, in which the reactant gives a product containing the same number of carbon atoms. Kekulé assumes that this will change peripheral atoms or groups but will leave the carbon skeleton (*"Kohlenskelett"*) unchanged.

The second proposes that in changes that disrupt the carbon skeleton itself, giving products with fewer numbers of carbons than the reactant, the atoms of the fragments obtained have the same mutual attachments internally as they had in the intact reactant. Thus, each fragment is a lower homolog of the reactant or is related to a compound that is so homologous by a simple transformation (for example, by replacement of hydrogen with oxygen). Then the fragments can be mentally recombined to permit us to deduce the structure of the original molecule. Both of the rules derive from the same idea, which came to be called the *principle of minimal structural change*: the carbon skeleton is like a sculptor's armature, onto which the chemist can append materials and transform them without concern that the underlying support will alter. It will be a major objective of this chapter to show

how this concept dominated the thinking of almost the entire chemical community and delayed the recognition of the phenomenon of molecular rearrangement for years.

We recognize in these two rules the entire basis of two great branches of organic chemistry. The first really provides, for example, the source of our confidence that we can undertake a multi-step synthesis of an organic compound with a strong expectation of arriving eventually at the correct structure. In a more general, pre-structural sense, a similar idea underlies the work of Dumas, Gerhardt, Laurent, and others on substitution reactions and the persistence of "types" in chemistry. [4]

The second rule forms the basis of the whole field of determination of molecular structure by degradative analysis. It tells us that we can deduce the structure of a complex molecule by examining the structures of its simpler decomposition products.

Of course, Kekulé was making a bold extrapolation from his examination of the small number of organic molecules then available. Furthermore, both of these rules rest upon an unarticulated assumption, which the philosophers call *ceteris paribus,* or "other things being equal." (see Chapter 6) It is an escape hatch which we hope will mitigate opprobrium should future events show that our theories or rules do not always hold.

Chemists today recognize that Kekulé's rules are not inviolate, but I think we may not be aware of the sequence of events after Kekulé's announcement of them, developments which eventually convinced the 19th-century chemists of that fact. The discoveries of the pinacol and benzilic acid rearrangements exemplify these refutations, which really constituted a kind of counter-revolution in chemistry. They came to light only about fifteen years after Kekulé's rules themselves. Although they occurred in two widely disparate fields and involved different sets of protagonists, their histories took strikingly parallel courses. Among the common features of these rearrangements is their designation by an allusive brief description which identifies the transformation, rather than by the name(s) of the discoverer(s). This is in con-

Figure 6.1 Rudolf Fittig, Edgar Fahs Smith Collection, University of Pennsylvania

trast to most other important (and many unimportant) new chemical reactions, which bear chemist's names. An inspection of why these two cases are special will tell us much about the dynamics of discovery at that point in the history of chemistry.

6.2
The Pinacol Rearrangement. Fittig's "Discovery"

The anonymity of the pinacol rearrangement is especially puzzling, since during much of the last century and continuing now, most reviews of it that attempt to trace its development come to the quite definite conclusion that the discoverer is known to have been Rudolf Fittig. Thus, a reviewer in 1938 writes: [5]

> "In 1859 Fittig published the results of a series of experiments which he had made in a study of the reducing action of sodium on acetone **1** (Scheme 6.1). The compound pinacol **2** was obtained. In the following year he published a paper [6] which showed that if this substance be heated with sulfuric acid a ketone, pinacolone **3**, is produced. Numerous examples of this reaction, now known as the pinacol rearrangement, have been found since the time of Fittig's discovery ..."

Scheme 6.1

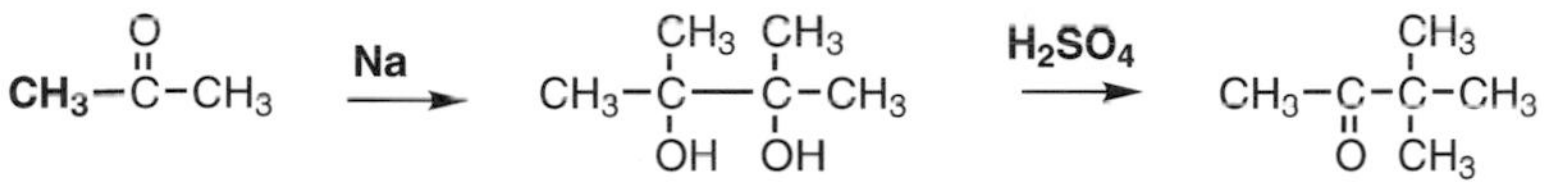

$CH_3-C(=O)-CH_3$ $\xrightarrow{Na}$ $CH_3-C(CH_3)(OH)-C(CH_3)(OH)-CH_3$ $\xrightarrow{H_2SO_4}$ $CH_3-C(=O)-C(CH_3)_2-CH_3$

acetone 1 **pinacol 2** **pinacolone 3**

We shall see that the form of these statements is misleading. Consider the word "study." The dictionary definition of "study," is an "act or process of acquiring by one's own efforts knowledge or information of a particular subject." By this criterion, we will be obliged to say that Fittig's experiments had entirely different objectives and were not knowingly directed toward "a study of the reducing action of sodium on acetone". He did not know the nature of the compound pinacol which he had produced, and specifically, he did not know it had been formed by a reduction. Further, he did not know that the substance pinacolone to which it was transformed by acid was a ketone. In fact, he did not even know the atomic composition of pinacol or pinacolone or, for that matter, of acetone. Finally, we must question whether the terminology, "Fittig's discovery", is appropriate. Toward the end of this essay, I argue that Fittig's work was a necessary part of the eventual full discovery but by itself fell well short of meriting that term.

These points were quite well understood in the late 19th century, for example, in textbook accounts of the history, [7–9] but somehow, they were forgotten later. Actually, we might ask, if Fittig merits the unequivocal credit given him by many authors, why has the chemical community withheld the accolade "Fittig rearrangement"? I cannot claim to have found the answer to this apparent contradiction, but I do believe that a (perhaps inadvertent) sort of rough justice is at work here. I intend to show that not only did Fittig not really discover the pinacol rearrangement, he *could not have done so at the time.* This history raises again the vexatious question of just what we mean by the word "discovery".

6.2.1 Early History of the Pinacol Rearrangement

Contrary to the implication in the frequently over-simplified portrayal of its history, the discovery of the pinacol rearrangement was not a sudden event. The process actually took about fifteen years, from 1859–1874. That was a crucial period in the history of organic chemistry, one which embodied what has been called a "quiet revolution" [4;10] in the whole conceptualization of the science. On the path to clarification of the pinacol rearrangement, delayed responses to the new ideas entangled some chemists in an impenetrable intellectual thicket.

6.2.2 The Atomic Weight Controversy and the Rise of Structural Theory

To examine the events that led to the emergence of the rearrangement, we will need to adopt temporarily the modes of thought of the organic chemists of that time. The unitiated reader of the early literature may find this is a daunting task. Before the advent of the structural theory in 1858, the basic principles of organic chemistry were a confused turmoil of conflicting ideas expressed in several different notations. Not only were the protagonists constantly in battle over their theories, but they sometimes abandoned their positions to join the opponents on one or another point. The discussion here is a mere survey, but thanks to the patience and scholarship of historians such as, *inter alia,* Ihde, [11] Rocke, [4] and Hjelt, [12] we are fortunate to have more thorough treatments, to which interested readers are referred.

By the middle of the 19th century, the chemical community faced a crisis over the issue of atomic weights. After several decades of dispute, many chemists had reached a consensus on the values of the atomic weights of the most common elements found in organic chemistry. These "conventional" weights, [13]; for example, H = 1, C = 6, and O = 8, by 1850 "had become nearly universally accepted." [14] The choice of atomic weights had profound consequences for chemistry. This is shown by comparing the molecular formulas for a few common substances derived on the basis of "conventional" weights with the modern formulas. Note that the "conventional" atomic weights lead to molecular formulas that look strange to the present-day chemist: Methane would be CH_2 or a higher multiple instead of CH_4, ethanol C_2H_3O or $(C_2H_3O)_n$ instead of C_2H_6O, etc. One way to think about the differences is to recognize that the modern atomic weights would require that for a given number of hydrogens, the number of carbons and oxygens each would have to be half the number in the "conventional" formulas. Also, note that water in the conventional system was OH or $(OH)_n$.

Berzelius, who in the early 19th century perhaps came closest to the embodiment of chemical authority, had developed a table of atomic weights [11] whose values were based upon a scale in which oxygen was set equal to 100. On this basis, he argued, following Dulong, that carbon should have the relative weight of 76.437. If we convert these proportionately to modern atomic weights, based on O = 15.9994, we get C = 12.25. He assigned weights to a number of the other elements which also were close to their modern values.

Arguments in favor of a set of "reformed" atomic weights (H = 1, C = 12, O = 16, etc.) similar but not identical to those of Berzelius were put forward by Dumas and especially by Gerhardt and Laurent. By the time of the famous Karlsruhe Conference in 1860, many chemists already had joined the "reform" camp, and most of the remaining ones did so after the conference. Cannizzaro's arguments presented there are often credited with having persuaded the holdouts, but there was another reason. [15] This was the rise of the structural theory, introduced just before then by Kekulé, Couper, Butlerov, and others. [10;11] Their analysis and representation of many organic structures depended on the assumption of the "reformed" atomic weights.

A central hypothesis of structural theory was the idea that each element had a characteristic "atomicity", or binding capacity, or as it was eventually called "valence". In the late 1850s, it became increasingly clear to the structuralists that no internally consistent set of structures for the known organic compounds could be devised using the "conventional" atomic weights. As Kekulé and his co-structuralists realized, to solve the problem of structure, it is essential to have *both* a set of correct weights and a set of correct valences. The "reformed" atomic weights, together with a set of valences C = 4, O = 2, Cl = 1, H = 1, etc., led in a straightforward manner to satisfactory structural formulas for known compounds and to verifiable predictions of new structures. Thus, self-consistency now required that the "reformed" atomic weights be accepted. In fact, many chemists were now quietly [15] converted to the new atomic weights perforce: if one wanted to apply the powerful new structural theory, one could hardly argue at the same time that the old atomic weights were correct. Keep in mind that the early events of the pinacol rearrangement story took place in 1858–1860 and thus came right at the time of a major discontinuity in the development of organic chemistry.

6.2.3
The Structure of Acetone. Uncertainty Piled on Obscurity

The organic chemistry of the first part of the 19th century dealt with a few natural products and a few commercial processes. It was not until the rise of the structural theory that the now familiar phenomenon of explosive growth in the number and complexity of organic chemicals occurred [4]. Fittig's work on the pinacol rearrangement grew out of an interest in a readily available substance, acetone.

Acetone then already was an old chemical, having been observed as early as 1595 as a product of the distillation of "sugar of lead" (lead acetate). [16] By the early 19th century, when it was known as "pyroacetic spirit," it had become an article of commerce, being produced by the dry distillation of metal acetates, especially calcium acetate, or also by the distillation of wood or other carbohydrate-rich materials with lime.

The first real scientific evidence on the nature of acetone came in 1832 from the work of Liebig, [17] who determined its elemental percentage composition by combustion analysis: C, 62.50, 61.94%; H, 10.47, 10.46%. [18] These percentage composition values, derived from the Berzelius atomic weights, of course fit the modern empirical formula of acetone $(C_3H_6O)_n$, where n is an integer. With the "reformed" atomic weights, with C = 12 instead of 12.25, the percentage values would be slightly different, but the same empirical formula would result.

However, by the "conventional" atomic weights of the mid-19th century, this composition would lead to the formula $(C_3H_3O)_n$. One must note that agreement with percentage composition alone is never a guarantee of atomic composition. Not only integral multiples but also a large number, in fact, an *infinity* of other possible molecular formulas, containing only C, H, and O, can qualify within the experimental error of ~0.3%. An early example comes from the work of the English chemist Robert Kane, [19] who obtained the hydrocarbon mesitylene from the reaction of the dimer (now known to be of the aldol type) af acetone, diacetone alcohol, with sulfuric acid. He assigned the molecular formula C_6H_8 on the basis of the percentage composition C 90.19, H 9.81 (Berzelius weights). The true atomic composition is C_9H_{12}, which has almost the same percentage composition: C 89.93, H 10.06.

It should be clear that those chemists who adhered to the "conventional" atomic weights suffered a disabling handicap in their attempts to determine molecular structure. The first requirement of that process must be to count the number and kind of atoms. The "conventional" atomic weights gave a completely erroneous measure of these properties, and those who used those weights thus had no chance ever to determine the structure of any compound.

During the years before and even after the Liebig analysis, although the structural theory lay years in the future, numerous attempts to determine the chemical nature of acetone were made. I mention only a few of these here.

Among the earliest investigators was the Irish Huguenot chemist Richard Chenevix. He believed that acetone belonged to the alcohol family and studied the action of metallic potassium on acetone in 1809 but obtained inconclusive results.

However, suggestive findings emerged in experiments on this reaction by Löwig and Weidmann in 1840. [20] They concluded that the main products were mesityl oxide and phorone (resulting from what we now would call aldol condensation), the same products as previously had been observed from the action of potassium hydroxide.

6.2.4
Fittig Attempts to Alkylate Acetone

Fittig started out with the modest objective of repeating the Löwig-Weidmann experiment with sodium instead of potasium. Superficially, one might have expected the two experiments to give similar results, but that was not to be the case. Fittig's serendipitous findings in this research were to take on unanticipated significance for all of organic chemistry. His observations were seminal, but he was not fated to carry through his initial findings to their startling conclusion. To understand the reasons for this, we must take a moment to learn something of the life and career of Fittig. This will allow us to analyze and perhaps identify the particular circumstances that diverted the development of the subject into new channels. It may also help us to understand why Fittig's allegiance to the "conventional" atomic weights kept him from solving this problem.

6.2.5
Rudolf Fittig (1835–1910)

From a lengthy obituary notice by Fittig's one-time student, Friedrich Fichter, [21] we learn that Wilhelm Rudolph Fittig was born in Hamburg in 1835, the son of a private school principal. As one of many children in the household, Fittig had to contribute to the family finances from an early age. At sixteen, he already was a teacher at the J.H. Hoffmann private school and also at a reform school. That Fittig was a talented and brilliant student was apparent early on, when despite the demands of his heavy work schedule, he passed his maturity examination with the highest honors.

In 1856, Fittig entered the University of Göttingen, where Friedrich Wöhler led the institute of chemistry. Fittig became associated there with Professor Heinrich Limpricht, who held the rank of extraordinary professor (roughly associate professor) and was the leader of a section of organic chemistry. Fittig carried out his early research on the sulfonation of benzamide and toluamide under Limpricht's supervision. The resulting paper [22] appeared in the *Annalen* in 1858 with Fittig as the sole author. It is significant that Limpricht was similarly generous with other students. [23]

Fittig went on to a noteworthy career as a professor of organic chemistry, becoming associate professor at Göttingen, moving then to Tübingen and later to Strasbourg. He and his students published almost 400 scientific papers, among which some of the best known were those reporting a modification of the Wurtz synthesis of alkanes to permit the synthesis of alkylarenes (now called the Wurtz-Fittig reaction). There is no doubt that during his time, he was one of the major figures in German organic chemistry.

6.2.6
The Acetone to Pinacol Reaction

When Fittig, still working in Limpricht's laboratory, started his work on acetone, he was following well-trodden paths. Nevertheless, he felt that much more was to be done. Referring to those early studies, Fittig remarks [24] (my translation):

> "Such an investigation at that time was so difficult and one could rely so little in general on the results obtained; yet it is noteworthy that even so, Chenevix already could have been been led to exclude acetone from among the alcohols had he been able to combine certain of its properties with those results."

The "certain of its properties" which Fittig had in mind seem to have been the atomic composition of acetone. His argument fails, however, because he himself was using the "conventional" atomic weights. Liebig's 1832 elemental analysis [17;18] of acetone, which used Berzelius's atomic weights, H = 1, C = 12.2, O = 16, gave the correct formula C_3H_6O, but Fittig, replacing these with "conventional" weights, H = 1, C = 6, O = 8, obtained thereby the incorrect formula C_3H_3O for acetone.

Among the experiments Fittig planned for the acetone-sodium reaction was a test of whether one could form and trap a sodium salt from acetone, what we would call today a sodium enolate. Only a year before, Ebersbach [25] had reported an analogous reaction with valeraldehyde and had carried out a successful alkylation of the sodium salt with ethyl iodide. The result of Fittig's experiment was disappointing in that he was unable to find the expected alkylation product. Instead, he found a mixture from which he isolated a quantity of large, tablet-shaped crystals, which he called "paraceton." This was the substance that later would be called "pinacon" and still later, "pinacol." As we shall see, the confusing changes of name were caused by misunderstandings of the actual nature of the substance.

The crystals contained hydrate water, and at first, the isolation of the water-free substance was difficult. Fittig subjected three samples of the hydrous crystals to combustion analysis and found average values of C = 32.6%, H = 11.5%. He interpreted these results with the molecular formula $C_6H_6O_2 \cdot 6$ aq. This formula requires C = 32.1%, H = 10.7% (Remember that in the "conventional" atomic weight system, water is represented as HO). Fittig's name for the substance, "paraceton", came from his belief that it was a *dimer* of acetone. It should be clear from this outline that, as I already have suggested, Fittig had no idea of carrying out "a study of the reducing action of sodium on acetone". [5] That was not his objective, nor in Fittig's mind was that the result. He started out trying to do an alkylation and in his view, succeeded instead in causing a dimerization, not a reduction.

6.2.7
Conflict with Städeler. A Pillow-Fight at Midnight in a Dark Cellar Between Blindfolded Combatants

Later in the same year, Städeler, [26] who also had been working on acetone for some time, published a paper challenging Fittig's results and interpretations. We should note that the argument between Städeler and Fittig, although rancorous, was inconclusive and doomed to eventual inconsequentiality, because both chemists used the "conventional" incorrect atomic weights. In the interest of brevity, I merely summarize it here.

Städeler's elemental analyses convinced him that pinacol was not a dimer of acetone but rather a reduction product of a dimer. For reasons that need not concern us here, it turned out that the difference between the percentage compositions required by Fittig's putative dimer and Städeler's reduced dimer were almost within the experimental error of the measurements. Eventually, it was shown by others that Städeler's conclusion that pinacol is formed from acetone by a reduction was correct. A fair historical treatment thus would be obliged to credit Städeler, not Fittig, with recognizing this.

6.2.8 The Transformation of Pinacol to Pinacolone

Fittig's second paper on this subject, published in 1860, contained a rebuttal of Städeler's arguments [6] but more significantly, reported the observation that baffled the chemical community for years afterward and ultimately generated a vast field of chemistry. He found that heating "paraceton" with aqueous sulfuric acid or hydrochloric acid gave rise to a new substance which Fittig called "pinacolin" (elsewhere rendered as "pinakolin" and now known as pinacolone). He assigned the formula $C_{12}H_{12}O_2$ to pinacolone. The reasoning in his discussion of what its nature might be was quite confusing, but in summary, he believed that pinacolone is an isomer or polymer of the anhydrous form of pinacol. His thinking here is muddled, since $(C_6H_6O_2)_2$ is $C_{12}H_{12}O_4$, not $C_{12}H_{12}O_2$.

The problem of representing the *structures* of pinacol and pinacolone remained. No matter how strenuously the two investigators contested the validity of each other's conclusions, their efforts could not bring further progress as long as they adhered to their custom of using the "conventional" atomic weights. I imagine that they gradually recognized that they had reached an impasse. As far as I have been able to determine, neither Fittig (after 1860) nor Städeler (after 1859) ever published another word about "paraceton" or "pinacon" or "pinacolin".

6.2.9 Loyalty and Dependency. Fittig's Atomic Weight Problem

Before we go on to later developments in this story, we must make clear Fittig's own attitude to the atomic weight issue. It is significant that by 1859, as Rocke has pointed out, Fittig was among the large number of chemists who were climbing onto the Kekulé bandwagon. Indeed, Rocke [27] already has called attention to a passage from Fittig's diary dated September 17, 1859, quoted by Fichter, [28] in which Fittig describes his profoundly epiphanic reaction to the appearance of the first fascicle of Kekulé's textbook:

> "I used the free time for the lectures on organic chemistry by Kekulé, of which the first part appeared a few weeks ago. The book is excellent in every respect, and in my opinion, all of the books on organic chemistry known to me cannot bear comparison with it. I read it page by page and with truly intense interest, and as was said of St. Louis, 'we are all in the true sense students of Kekulé. The field that has been sown with his ideas will be cultivated everywhere and will bear unceasingly the richest harvest.'"

Kekulé had made a major point, both in his paper on the structural theory and in his textbook, that the "conventional" atomic weights must now be abandoned in favor of the "reformed" weights. Fittig's reaction to the book in September, 1859 was an emotional experience almost as intense as a religious conversion. How then are we to understand his adherence to the discredited "conventional" atomic weights in his sec-

Figure 6.2 Charles Friedel, Edgar Fahs Smith Collection, University of Pennsylvania

ond paper on pinacol, [6] which was submitted to the *Annalen* several months later, in December, 1859?

The answer to this question must be speculative. It does not seen highly probable that Fittig, despite all his enthusiasm for Kekulé's book, had not yet absorbed its lessons. Fittig, after all, was a highly motivated, conscientious, thorough chemist. It seems to me unlikely that he would absent-mindedly overlook the contradiction between his own line of thinking on constitutional questions and that of Kekulé, which had made such a deep impression on him.

I believe a more plausible conjecture is that there were forces external to the chemistry itself that kept him temporarily in thrall to the "conventional" atomic weights. What might these have been?

Consider the *mise en scène:* In 1859–60, the 24-year old Fittig was still dependent on Limpricht for laboratory space. This may be seen from the headings of Fittig's two papers on pinacol, which were headed "Researches from the Academic Laboratory in Göttingen; Section of Prof. H. Limpricht". Moreover, Fittig was sensitive about his personal relationship to Limpricht. From Fittig's diary [29] we learn that Limpricht had diverted Fittig from his own research to help with Limpricht's study of the sulfonation of benzoic acid. Fittig was apprehensive about what this would mean to the progress of his own career. I think it is likely, for reasons to be examined below, that Fittig's concern went beyond the irksome delay to be endured while he worked on a project not of primary interest to him. He was so reluctant to work on the sulfonation problem that he even considered leaving Limpricht's laboratory and asking Wöhler for a place. Although the matter was smoothed over when Limpricht eventually allowed Fittig to publish his results from the sulfonation study independently, one gets the impression that Fittig was wary about putting himself at cross-purposes with his mentor.

One might wonder whether Fittig was being over-cautious here, since Limpricht had been quite generous in granting independent publication to other students. [23;30] Nevertheless, there can be no doubt that Limpricht's name and laboratory would be closely associated with these studies on sulfonation of benzoic derivatives. Limpricht had been engaged in them for some time, and they were a centerpiece of his research program. It is crucial to know that his interpretation of the results employed the "conventional" atomic weights, as was usual for him.

What could have hardly escaped Fittig's notice was that at just about that time, Kekulé had begun a strong assault on the basic approach of Limpricht's project on the sulfonation of benzoic derivatives. As it happened, Kekulé had published the first part of his structural theory in 1857, [31] and in developing his arguments on the quadrivalency of carbon, he pointed to Limpricht's immediately preceding work [23;30] on the sulfonated arenecarboxylic acids as an example of the erroneous nature of the "conventional" atomic weights. In fact, in Rocke's words, [32]

> "These two papers [of Limpricht] gave Kekulé the opportunity he had been looking for to publish his structural ideas, for Limpricht had opened himself to criticism in several ways, and such a critique provided the rhetorical launching platform for more general considerations."

Although Limpricht published a rebuttal, [33] Kekulé soon followed with an expanded counter-rebuttal in the climactic paper [3] of his structural theory. Limpricht thus found himself battling Kekulé, in a very public arena, over the most significant question in organic chemistry.

Fittig now could not have been unaware of his own vulnerability. Although by his own words, we know that he was attracted to Kekulé's ideas, he also could hardly have ignored the fact that adopting them publicly at that moment would signify a break with his mentor, one which might have been considered an act of disloyalty. For someone in his dependent position, this would have been a risky step indeed. At the same time, as an ambitious and perspicacious young chemist, he would have been wary of an association with what could well turn out to be a prominent defeat for Limpricht. One can only sympathize with his dilemma.

So is it possible that Fittig somehow rationalized and suppressed his own doubts about the "conventional" atomic weights when he published his 1860 paper [6] on pinacol? Did loyalty and dependency win out over science? I have no evidence on this point, but it is noteworthy that within the following year, Fittig [34] published a paper on another subject in which he once and for all abandoned the "conventional" atomic weights and adopted the "reformed" values without dissent and without even a mention that this step represented a complete turnover in his chemical thinking. In this, he joined many other organic chemists of the time, who quietly became adherents of Kekulé's ideas. [10;15] This paper, Fittig's first truly independent publication, no longer bears a heading giving its origin as the "Section of Prof. H. Limpricht." By then, Fittig had passed his habilitation and apparently had been taken on as an assistant to Wöhler. Whether this change in academic status finally facilitated Fittig's scientific switch is admittedly a matter of conjecture, but I think we cannot ignore the possibility that Fittig's awkward personal circumstances at a crucial time may have prevented him from thinking the pinacol problem through to a satisfactory answer.

6.2.10
The Structures of Pinacol and Pinacolone. Friedel and Finally Butlerov

On the other hand, the advocates of "reformed" atomic weights did not immediately solve the problem either. Foremost among these was August Kekulé, then a professor at Ghent. Kekulé already had published the main outlines of his ground-breaking structural theory of organic chemistry in the famous 1858 paper *"On the Constitution and Transformation of Chemical Compounds and on the Chemical Nature of Carbon."* [3] As already has been mentioned, the theory depended critically on the "reformed" atomic weights. Kekulé was clearly aware of the pinacol problem, and in fact, the next discussion of Fittig's "paraceton" in the literature seems to have come in Kekulé's textbook, published in a series of fascicles starting in 1859 and in bound form in 1861. [35] In the disagreement between Fittig and Städeler, Kekulé's text favors the general idea of Städeler's interpretation of the acetone → pinacol reaction as a reduction, although his overall concept (reformed) differs from Städeler's (conventional) in that Kekulé uses C_3H_6O for acetone, and $C_6H_{14}O_2$ for pinacol. Kekulé's empirical formulas were correct, but as we shall see, many years were to pass before the structural significance of these would emerge.

Soon after these events, Charles Friedel, [36] at the National School of Mines in Paris, was able to improve the yield of pinacol in the reduction of acetone by substituting sodium amalgam for sodium, following a method introduced by his former mentor, Wurtz. Moreover, Friedel succeeded in preparing anhydrous pinacol, which with "reformed" atomic weights gave the molecular formula $C_6H_{14}O_2$, a reduction product of acetone as had been advocated by Stadeler and by Kekulé. As for the structure, Friedel was able to suggest at this time (1862) only that the substance behaved like a glycol in its reactions. He was unable to convert it to the corresponding diacetate, which would have amounted to a convincing confirmation of his hypothesis.

The next important work on pinacol (pinacon) was reported three years later by Linnemann, working in Lemberg. [37] He confirmed Friedel's preparation of anhydrous pinacol as well as its elemental composition, $C_6H_{14}O_2$, and its conversion to several different hydrated forms, which also had been reported by Fittig. Linnemann isolated both a crystalline and an oily form of anhydrous pinacol and believed the two were "isomers," an assignment that later proved incorrect. Linnemann's most telling experiment was the oxidation of pinacol with chromic acid, which smoothly converted it to acetone. He carried out a similar oxidation of benzpinacol to benzophenone. Strikingly, he was unable to deduce a structural formula for pinacol from these data. His paper concludes that little had yet been established about the chemical nature of the pinacols, and he expresses the hope that Friedel would solve the problem:

> "If I am not mistaken, one can expect a closer examination of pinacol by Friedel, so that the chemical nature of this interesting ... substance hopefully soon will be clarified."

Linnemann was being overly modest here. His results showed two new and important things: first, that the formation of pinacol from acetone was indeed a reduction, and second, that pinacol must have a structure in which two oxygenated carbons were connected directly to each other. This obviously goes a long way toward solving the mystery of the pinacol structure.

In fact, even the partial fulfillment of his stated hope that Friedel would solve the problem was delayed for several years. It was not until 1869 that Friedel published a review [38] on the properties of aldehydes and ketones which contained an extensive section on pinacol. With a minimum of experimental evidence, he deduced the correct structure of pinacol (pinacone) as **2**. His major argument was an intuitive guess, remarkable for the time, about the mechanism of the reduction of acetone. He visualized pinacol as the result of "addition of one atom of hydrogen to one molecule of acetone, followed by dimerization of the monovalent radical so produced." He claimed that his structure of pinacol as a glycol was confirmed by some chlorination experiments. In fact, the best that can be said of them was that they did not refute the structure **2**.

Curiously, in the 1869 paper, Friedel does not mention the much stronger evidence brought forward by Linnemann's oxidation of pinacol to acetone [37] several years before, which we have already described. This made the pinacol structure **2** highly likely. Friedel does mention this work in his 1876 review in Wurtz's *Dictionnaire*, [7] but even then, it appears only as an isolated fact, without comment on its significance. On the other hand, he goes to some length in that paper to criticize Linnemann for the relatively trivial mistake of calling the crystalline and oily forms of anhydrous pinacol "isomers." In fact the oil is simply a supercooled specimen of the same compound as the crystalline material and can be converted to it just by agitation with a glass rod.

Perhaps the most important contribution of Friedel's 1869 paper [38] was the demonstration that the acid-induced transformation of pinacol to pinacolone is an overall dehydration. This became immediately apparent from the molecular formulas. Friedel proposed that the two compounds are related in the same way that ethylene glycol **5** is related to its dehydration product, ethylene oxide (**6**, Scheme 6.2). If pinacol is **2**, by this analogy, then pinacolone is tetramethylethylene oxide, **4**.

Friedel did not report direct experimental studies on the strucure of pinacolone (pinacoline) until 1873, in a series of reports in collaboration with R. Silva [39–46] These were not papers in the usual sense but brief accounts, given by Friedel to meetings of the Chemical Society of France in Paris and published both in the Society's *Bulletin* and in the *Berichte* without references to the literature or experimental details. The results also were reported, together with later findings, in Friedel's 1876 review article. [7]

Only one of these experiments eventually proved to be of significance to the structural problem. This was the finding that pinacolone can be oxidized by dichromate to CO_2 and an acid isomeric with valeric acid. Friedel and Silva at first proposed [41] that this acid, "pinacolinic acid," in contrast to other known organic acids, did not have the usual carboxyl group but instead contained a hydroxyoxirane group, as in **7** (Scheme 6.3). (They actually show an oxirane with H in place of the OH in **7** (Scheme

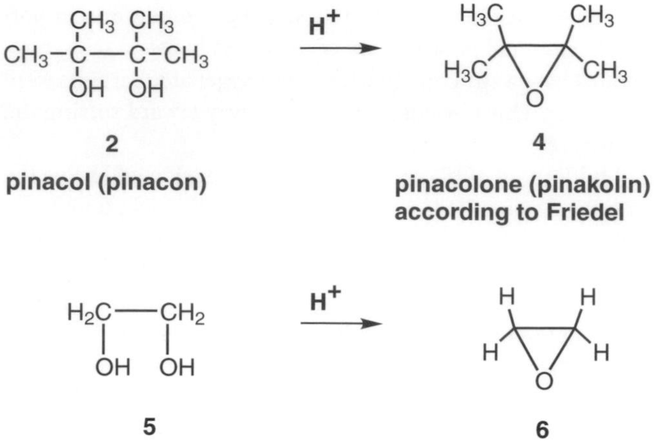

2 pinacol (pinacon) → (H^+) 4 pinacolone (pinakolin) according to Friedel

5 → (H^+) 6

Scheme 6.2

6.3), but this must be a typographical error, since that would give a compound with the wrong empirical formula). They thought they had found an analogy for the unorthodox oxirane hypothesis in the claim of Grimaux to have shown a similar structure for the known compound benzilic acid. This eventually proved to be incorrect.

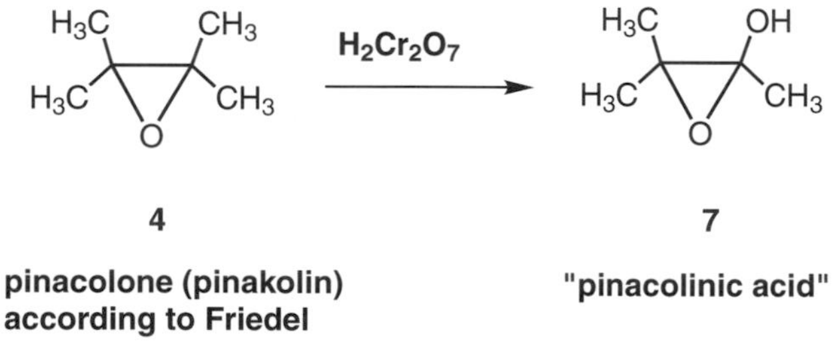

4 pinacolone (pinakolin) according to Friedel → ($H_2Cr_2O_7$) 7 "pinacolinic acid"

Scheme 6.3

It should be clear that Friedel's difficulty with the structure was the result of his inability to free himself from his adherence to the rule of minimal structural change. Dehydration of pinacol to an oxirane structure for pinacolone not only provided a way of accommodating the change in atomic composition but also met his assumed obligation to preserve the rule of structural invariance of the carbon skeleton. However, this left him no way to account for the acidity of pinacolinic acid, the oxidation product of the oxirane pinacolone. Hence the startling proposal that pinacolinic acid is a hydroxyoxirane.

The fog surrounding all this chemistry was finally disspelled by two brilliant papers of Alexander Butlerov of St. Petersburg. In the first of these, [47] published shortly after the first four papers [39–42] of the 1873 Friedel series, Butlerov accepted the pinacol structure **2** favored by Friedel and Silva, but he questioned the necessity for their special hypothesis of a unique new kind of acid structure for the oxidation product $C_5H_{10}O_2$ obtained from pinacolone. He proposed instead that Friedel and Silva's "pinacolinic acid" was in fact trimethylacetic acid **8** (Scheme 6.4), an assignment which he supported by direct synthesis of that compound by the route shown

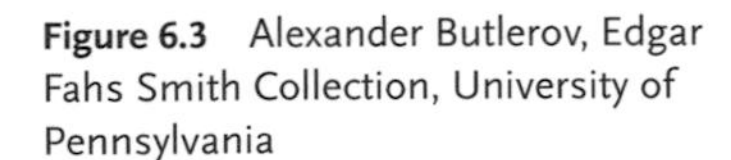

Figure 6.3 Alexander Butlerov, Edgar Fahs Smith Collection, University of Pennsylvania

in Scheme 6.4, namely formation of trimethylacetonitrile **9** from *t*-butyl iodide **10** and mercuric cyanide **11**, followed by hydrolysis to trimethylacetic acid **8**. Butlerov noted that his synthetic acid showed properties very similar to those of the "pinacolinic acid" obtained by Friedel and Silva from pinacolone. Later, in a fifth paper of their series, the latter authors agree that their acid is "probably identical with trimethylacetic acid."

Scheme 6.4

$$\underset{\mathbf{10}}{(H_3C)_3C{-}I} \xrightarrow[\mathbf{11}]{Hg(CN)_2} \underset{\mathbf{9}}{(H_3C)_3C{-}CN} \longrightarrow \underset{\mathbf{8}}{(H_3C)_3C{-}CO_2H}$$

After having established to his own satisfaction the identity of the acid, Butlerov [47] forged ahead boldly to suggest that the formation of pinacolone from pinacol in acidic solution involves a *molecular rearrangement of the carbon skeleton* and dehydration (Scheme 6.5) to form pinacolone, which is simply methyl *t*-butyl ketone **3**. He pointed out that such a reaction finds analogy in the already known transformation of ethylene glycol **5** to acetaldehyde **12**.

Friedel and Silva [46] disagreed, maintaining that the rearrangement must take place during the oxidation of pinacolone to trimethylacetic acid, since it is "not probable that so gentle a reaction as the hydrogenation of acetone brings about a rearrangement of the atoms, whereas that can be assumed during an oxidation." They give no justification for either of these assumptions, but in any case, it is clear that their reasoning has gone completely off the track here. Butlerov never claimed that

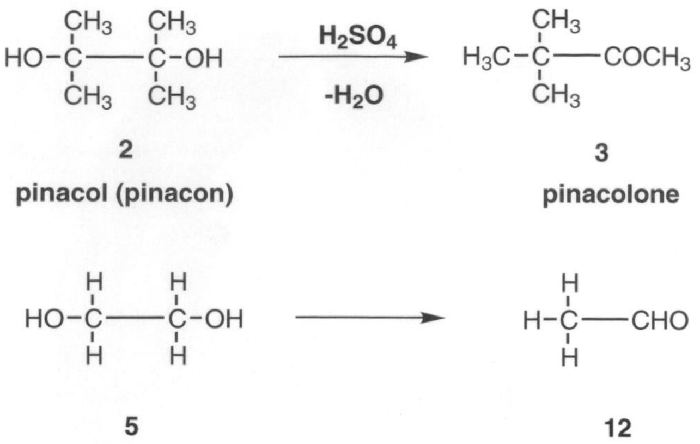

$$\mathrm{HO{-}C(CH_3)_2{-}C(CH_3)_2{-}OH} \xrightarrow[-H_2O]{H_2SO_4} \mathrm{H_3C{-}C(CH_3)_2{-}COCH_3}$$

2 pinacol (pinacon) → 3 pinacolone

$$\mathrm{HO{-}CH_2{-}CH_2{-}OH} \longrightarrow \mathrm{H{-}CH_2{-}CHO}$$

5 → 12

Scheme 6.5

the rearrangement occurred in the reduction of acetone. In fact, he agreed with Friedel and Silva on the structure of pinacol as an unrearranged dimeric reduction product of acetone. Butlerov's suggestion was that *the rearrangement occurred during the acid treatment required for the pinacol → pinacolone reaction.*

Butlerov proposed [47] to confirm his proposal by the synthesis of pinacolone itself **3** from trimethylacetyl chloride **13**, a task he completed the following year [48] by reaction with dimethylzinc **14** (Scheme 6.6). He considered the proof of structure of pinacolone then to be complete.

To my knowledge, Friedel himself never responded to Butlerov's synthesis. Apparently, both he and Butlerov turned the remaining problems over to students. Pawlow [49] in Butlerov's laboratory provided confirmatory evidence that pinacol is tetramethylethylene glycol, and much later, Friedel's student Couturier [50] confirmed Butlerov's structure for pinacolone.

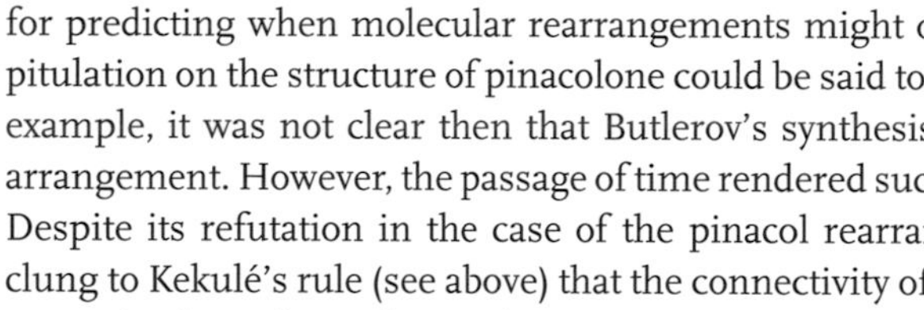

$$\mathrm{H_3C{-}C(CH_3)_2{-}COCl} \xrightarrow[14]{(CH_3)_2Zn} \mathrm{H_3C{-}C(CH_3)_2{-}COCH_3}$$

13 → 3

Scheme 6.6

One might argue that in 1874, at a time when chemists did not have reliable means for predicting when molecular rearrangements might occur, Friedel's apparent capitulation on the structure of pinacolone could be said to be premature. Logically, for example, it was not clear then that Butlerov's synthesis did not itself involve a rearrangement. However, the passage of time rendered such doubts less and less likely. Despite its refutation in the case of the pinacol rearrangement, organic chemists clung to Kekulé's rule (see above) that the connectivity of the carbon skeleton tended to persist throughout chemical reaction.

Butlerov thus recognized one of the first carbon-skeletal rearrangements and, in my opinion, deserves credit at least equal to that later assigned largely to Fittig as the discoverer of the pinacol rearangement. Although Roscoe and Schorlemmer's textbook [8;9] of 1881 gives a clear account of the importance of Butlerov's work, I have been unable to determine the reasons why 20th century authors slight it or overlook it entirely. In some cases, one might suspect that simple lack of diligence is to blame. In others, however, this cannot be the reason. For example, the brief history in the 1961 textbook of Fieser and Fieser [51] shows that the authors were well aware of Butlerov's work but put a strangely dismissive twist on it:

> "The structure [of pinacolone] was established at St. Petersburg in 1873 by Butleroff, who seemed satisfied with a suggested analogy to the conversion of ethylene glycol **5** to acetaldehyde **12** (see Scheme 6.5). Actually the latter reaction involves merely dehydration to acetaldehyde enol **15** (Scheme 6.7), whereas the formation of pinacolone involves the migration of a methyl group under acid catalysis."

$$H_2C(OH){-}CH_2(OH) \;(\mathbf{5}) \longrightarrow H_2C{=}CHOH \;(\mathbf{15}) \longrightarrow H_3C{-}C(H){=}O \;(\mathbf{12})$$

Scheme 6.7

I digress a bit to examine this criticism. The Fiesers pass over Butlerov's insightful clarification of a whole field of controversy but choose to quibble with him about the mechanism of the putatively similar rearrangement of another compound. The enol mechanism they propose involves loss of a β–hydrogen and hence would be intermolecular. One might entertain it as a plausible alternative, but the basis for their conviction that the ethylene glycol rearrangement *must* go by that pathway is unclear. They give no documentation, and as far as I can determine, no definitive experiments on the mechanism of the ethylene glycol rearrangement have been reported. The enol mechanism had been proposed much earlier [52;53] and accounted for the formation of racemic product in the dehydration of certain enantiomerically enriched substituted 1,2-glycols. However, in another example, rearrangement of hydrogen occurs with some retention of enantiomeric excess, [54] a strong argument against the enol mechanism in this case and in favor of an intramolecular hydrogen shift mechanism, as proposed by Butlerov. Similar results favoring the intramolecular mechanism are found in the rearrangements of deuterium-labeled substituted 1,2-glycols. [55–57] For ethylene glycol itself, in the absence of experimental data, we have only computational results, which show that the internal hydrogen migration mechanism is at least a plausible pathway [58;59] and probably the preferred one. [60] Thus, for the time being, it is a fair conclusion that Butlerov's analogy between the pinacol and ethylene glycol rearrangements may well be valid.

6.3
The Benzilic Acid Rearrangement

6.3.1
Liebig vs. Laurent: A Fortuitous Observation and a Questionable Criticism

Almost simultaneously with the resolution of the last structural problems of the pinacol rearrangement, there was a parallel series of developments in the understanding of what later became known as the benzilic acid rearrangement. The personnel of the two fields overlapped only slightly. In fact, the literature of the time gives no indication that the two groups were even aware of each other's activity, of the mutual significance of their studies, or of the implications of both for chemistry in general. In this case, as in many others, it is as though the developments of chemistry took place in orthogonal domains of intellectual space. Only later could one see the relationships and realize how the workers at the time might have benefited if only they had known what their neighbors were doing. Again, as in the pinacol rearrangement story, the inability to imagine reactions in which the rule of minimal structural change is violated delayed the elucidation of the process for many years.

The benzilic acid story begins in 1838 with a corrosive assault by Liebig [61] on certain proposals of Laurent. We need not tarry over the details of the battles between Liebig and the French type-theorists, which have been discussed elsewhere. [4] It suffices to relate that Laurent had reported [62] that the action of alkali on benzil (then called "benzoyl" and shown in Scheme 6.8 in its present-day formulation **16**) gave benzoic acid ($PhCO_2H$) and an oil which he thought might be benzaldehyde (PhCHO). He proceeded to erect some rather far-reaching theoretical conclusions based upon these claims.

Liebig, who was involved in a series of running battles with the French school, was by then becoming exasperated with theoretical speculations, and in fact, was to abandon theory entirely in the following year. [4] He was convinced that Laurent's experiments were incorrect and undertook to repeat them. In the event, Liebig's concerns seemed at first to be justified, as he reported with a certain satisfaction:

> "There is no greater misfortune for a chemist than to be unable to tear himself loose from preconceived ideas, than to have his intellect so imprisoned that he creates an explanation for every contradictory finding, an explanation that is not derived from experiment ... My goal has been to prove to Mr. Laurent that his preconceived opinions have led him into error, and that goal has been reached."

What Liebig found was that the reaction of benzil with ethanolic potassium hydroxide gave neither benzaldehyde nor benzoic acid, the products Laurent had reported, but instead gave a new acidic substance, which according to Liebig's analysis led to the molecular formula $C_{28}H_{24}O_6$, just double the one we accept today, $C_{14}H_{12}O_3$. Liebig did not offer a trivial name for this acid, but in the following year, Zinin, [63] with Liebig's permission, worked out a modified method for its isolation from the reaction mixture and suggested the name benzilic acid. Scheme 6.8 shows the modern structures for benzil **16** and benzilic acid **19**, as well as their precursors benzoin

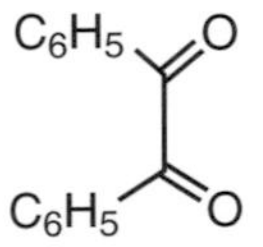

$C_6H_5COCOC_6H_5$ — **16** **benzil**

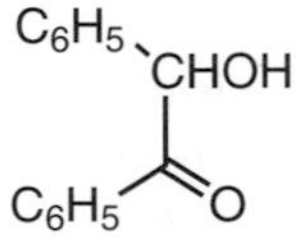

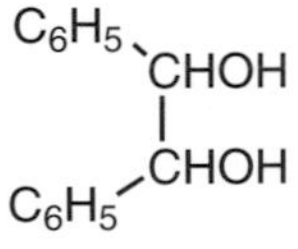

$C_6H_5CHOHCOC_6H_5$ — **17** **benzoin**

$C_6H_5CHOHCHOHC_6H_5$ — **18** **hydrobenzoin**

$(C_6H_5)_2C(OH)CO_2H$ — **19** **benzilic acid**

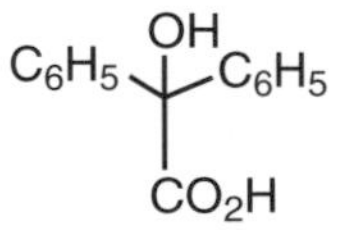

Scheme 6.8

17 and hydrobenzoin **18**, although as we shall see, many years were to pass before these were established.

It is ironic that Liebig's disagreement with Laurent may have been without any real experimental foundation. Neither Liebig nor Zinin explored the effect of conditions, particularly the concentration of alkali, on the formation of benzilic acid, but later workers [64;65] reported that the yield of benzilic acid decreases with increasing initial ratio of alkali to benzil, and at sufficiently high ratio, the benzilic acid yield becomes negligible. Under these conditions, the major product is said to be benzoic acid, together with a considerable quantity of tarry material. That the reaction course is indeed very sensitive to reaction conditions was later confirmed by Lachman. [66] The products obtained from benzil and alkali have been reported by several investigators to be complex and to vary with conditions. Thus, Liebig's criticism of Laurent may have been unjustified.

It now should be clear that Liebig's finding, which was the necessary first step in the discovery, was a classically accidental observation. Just fortuitously, Laurent apparently had chosen the conditions that favor benzoic acid formation and so had failed to observe the benzilic acid product. Liebig, for some reason chose a different set of conditions, which just happened to be the ones needed to produce benzilic acid in significant amounts.

6.3.2
The Benzil-Benzilic Acid Series. Futile Attempts to Assign a Common Skeletal Structure

Just as was the case in the rearrangement of pinacol to pinacolone, the rearrangement at the heart of the structural problem in the benzil-benzilic acid series, could not have been deduced before the development of the structural theory, which really did not begin until about 1858. [3] By 1869, Grimaux [67;68] in Paris as well as Limpricht and Schwanert, [64;69] and Jena [70] (in Limpricht's laboratory) in Greifswald had taken up the problem. Indeed, Kekulé [71] himself also had entered the free-for-all.

By 1869, chemical interconversions had been established among benzil, benzilic acid, benzoin, and hydrobenzoin. Opinion was divided as to whether these substances belonged to the 1,2-diphenylethane series (as in **20**, **16**, and **21**), or the 1,1-diphenylethane series (as in **22**, **23**, and **19**, (Scheme 6.9). For years, none of the investigators in the field gave any indication that they had ever considered the possibility that the carbon skeletons of these molecules might not all be the same, that is, that *both* the 1,1- and 1,2-diphenyl types of structure might be involved.

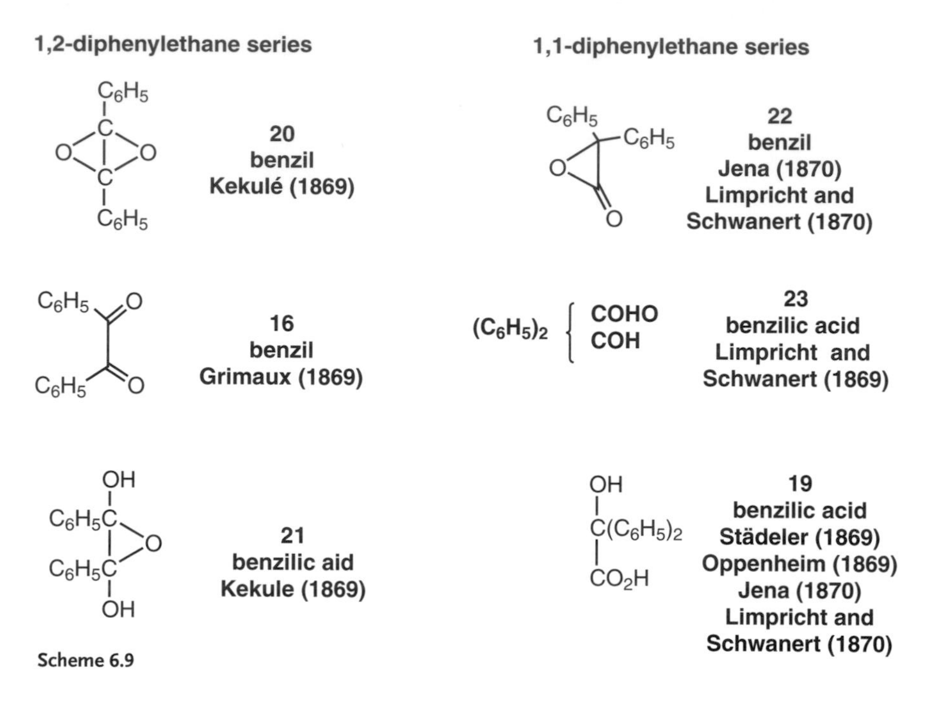

Scheme 6.9

Formula **23**, (Scheme 6.9) Limpricht and Schwanert's [64] tentative 1869 approach to the representation of benzilic acid, would be considered ambiguous today, although later, in their 1870 paper, [69] they implied that by **23** they really had in mind **19**. The year before, a paper by Grimaux [68] was presented to a meeting of the German Chemical Society by A. Oppenheim. Grimaux was puzzled by formula **23** and criticized it as

> "... inadmissible. It cannot be solved (*'aufgelöst'*), and it can contain the characteristic CO_2H carboxylic group only if a phenyl group has an arrangement entirely different from that usually assumed."

In the context of the rest of Grimaux's paper, this can only be interpreted to mean that he thought (correctly) that one could not have all the carbons of benzilic acid represented in a 1,2-diphenylethane skeleton (Ph-C-C-Ph) and still accommodate a carboxyl group. So there he was, on the very brink of realizing the true solution of the problem, and still not able to make the leap. Interestingly, Oppenheim, who had only presented Grimaux's work but had not been involved in it, added a footnote to the paper [72] which pointed out that the problem *could* be resolved with the formula **19**, in which the carbon skeleton of benzilic acid is assmed to be 1,1-diphenylethane Ph_2C-C. However, he did not go so far as to suggest explicitly that the formation of benzilic acid from benzil actually involves a rearrangement.

As we have seen, eventually Limpricht and Schwanert [69] replaced their original formula **23** for benzilic acid and adopted the Oppenheim [72]-Städeler [73] structure **19**. The 1,1-diphenylethane skeletal structure of **19** received further support from the work of Jena, [70] who showed that benzilic acid could be deoxygenated to diphenylacetic acid, Ph_2CHCO_2H, with hydroiodic acid, and could be oxidized to benzophenone, Ph_2CO, with dichromate.

Nevertheless, even after these developments, Limpricht, Schwanert, and Jena continued to adhere to the idea that benzil and benzilic acid both belonged to the same skeletal series, which in view of the events just described, they now believed to be the 1,1-diphenylethane series. This incorrect assumption, of course, was what led Jena [70] to propose the α-lactone structure **22** for benzil. Structure **22** still has the same carbon skeleton as benzilic acid **19** and differs from it only by the elimination of a molecule of water. In fact, in an attempt to show that benzil is the lactone of benzilic acid, Jena actually tried to carry out the dehydration of the acid with phosphoric anhydride, in other words, to reverse what he believed to be the course of the benzil-benzilic acid transformation. He observed only oily, unidentified products but optimistically pointed out that authentic benzil, prepared from benzoin or hydrobenzoin, sometimes fails to crystallize unless quite pure! The fact that he even published this negative result suggests that he remained convinced that the relationship of benzilic acid and benzil as **19** and **22** was correct. It seems clear that at this stage, the mental barrier to recognition that benzil and benzilic acid have different carbon skeletons was insurmountable, not only for Jena but also for his mentor Limpricht, who by 1870 [69] also had adopted Jena's structures.

6.3.3
The Rearrangement. Zincke Sees It But Can't Believe It. Baeyer Gets It

In the following year, 1871, Zincke [74] at Bonn began to express doubts that benzil and benzilic acid could belong to the same skeletal series. He showed that stilbene (Ph-CH=CH-Ph) and benzoin [Ph-CH(OH)-CO-Ph] can be oxidized with dichromate to give benzaldehyde (PhCHO) and benzoic acid ($PhCO_2H$), and that benzil gives benzoic acid. Zincke correctly interpreted these results as consistent with the presence of the 1,2-diphenylethane skeleton in each of these compounds. But he also knew, of course, of Jena's prior demonstration that benzilic acid upon oxidation gives *benzophenone* (Ph_2CO), not benzoic acid or benzaldehyde. Zincke comments [74]

> "The facts so far lead to the astonishing result that benzil, benzoin, and benzilic acid do not have the generally assumed relationship. Benzoin and benzil cannot be [as Limpricht, Schwanert, and Jena had proposed] the aldehyde and the anhydride of benzilic acid ... The transformation of benzil into benzilic acid, cannot be expressed by a simple equation, and very probably the formation of the

acid cannot occur without the occurrence of intermediate products, as has already been signified by the remarkable changes of color when benzil is heated with potassium hydroxide."

What a disappointment it is to read that passage! In the first part of the last sentence quoted above, Zincke seems at first to be headed directly toward realizing the occurrence of rearrangement, but at the last split-second, he veers off into fruitless speculation about "intermediate products."

Both Jena [65] and Zincke [74] recognized the importance of an independent synthesis of the structure Ph-CO-CO-Ph, dibenzoyl, which is the structure we accept today for benzil. In fact, Brigel [75] had reported such a synthesis by the action of sodium on benzoyl chloride, PhCOCl, in anhydrous ether, but neither of the later workers was able to repeat this result.

Matters stood at this point until 1874, when Symons and Zincke [76] synthesized benzilic acid from diphenylacetic acid (Ph_2CHCO_2H). They recognized that this left little doubt that benzilic acid has the 1,1-diphenylethane skeleton, with both phenyl groups attached to the same carbon atom. What is astonishing to the modern reader is that although they were able to draw the conclusion, which by now was unavoidable, that if the structure of benzil is **16**, then a molecular rearrangement of the carbon skeleton must occur in the reaction benzil → benzilic acid, they nevertheless still timidly edged away from confronting the issue by raising new and unfounded doubts about the correctness of structure **16** for benzil:

"... as long as *the constitution of benzil itself is still uncertain*, it appears to be idle to put forward theoretical speculations on the formation of benzilic acid."

One can hardly avoid the memory of Sherlock Holmes's admonition to Dr. Watson: [77]

"How often have I said to you that when you have eliminated the impossible, whatever remains, however improbable, must be the truth."

Thus, after five years of experimentation and argument by several laboratories, there was still no one eager to grasp the nettle of rearrangement. A quiet period ensued, and the next significant development seems to have been a paper by Baeyer [78] in 1877. In collaboration with his student Paul Friedländer, Baeyer studied a reaction that had earlier been observed by Heinrich Caro. Caro had studied the system only in a preliminary way and apparently never published his results but communicated them personally to Baeyer with permission to pursue the matter. Caro had found that heating phenanthrenequinone **24** (Scheme 6.10) with sodium hydroxide followed by neutralization caused the formation of a new crystalline acid. Baeyer and Friedländer determined the formula of this substance, $C_{14}H_{10}O_3$, which empirically results from addition of one molecule of water to the quinone. The structure was established by oxidation as that of diphenyleneglycolic acid **25**, which gave fluorenone **26** in good yield. Baeyer concludes:

"Accordingly it is probable that phenanthrenequinone **24** on heating with sodium hydroxide experiences a transformation analogous to that of benzil to benzilic acid. Addition of water and molecular rearrangement similarly would give the corresponding diphenyleneglycolic acid **25** from phenanthrenequinone **24**."

1. NaOH
2. H_3O^+

OH
CO_2H

$Cr_2O_7^=$

24 25 26 Scheme 6.10

Obviously, Baeyer had progressed beyond any lingering doubts over the formulation of the benzil → benzilic acid reaction and had simply swept aside the hesitations of Zincke and the others. Whether he was emboldened to do this by the prior appearance of Butlerov's papers on the pinacol rearrangement is not clear. Baeyer's paper gives no such reference, but by this time Butlerov's work had been in the literature for 3–4 years, and we may speculate that its impact on the community had been significant.

6.4
What Is a Discovery?

In the context of science, there is a perception (or misperception) that a discovery takes place in a sharply defined moment of time. I believe that this idea underlies many of the frequent disputes in the scientific literature over priority of discovery. Historians of science are well acquainted with this problem and tend to take a more nuanced view of the process, as in this passage from Thomas Kuhn: [79]

> "The sentence 'Oxygen was discovered,' misleads by suggesting that discovering something is a simple act assimilable to our usual ... concept of seeing. That is why we so readily assume that discovering, like seeing or touching, should be unequivocally attributable to an individual and to a moment in time. But the latter attribution is always impossible, and the former often is as well. ... [W]e can safely say that oxygen had not been discovered before 1774, and we would probably also say that it had been discovered by 1777 or shortly thereafter. But within those limits or others like them, any attempt to date the discovery must inevitably be arbitrary because discovering a new sort of phenomenon is necessarily a complex event, one which involves recognizing both *that* something is and *what* it is."

In the discovery of the pinacol rearrangement, fourteen years elapsed before Butlerov explained Fittig's transformation of pinacol to pinacolone. Applying Kuhn's scheme, we can say Fittig found *that* a reaction of pinacol occurred in acidic medium, and it was Butlerov who recognized *what* that reaction was. Similarly, Liebig discovered the existence of the transformation benzil → benzilic acid in 1838, some 22 years before Fittig's discovery of the transformation pinacol → pinacolone. The benzilic acid rearrangement is sometimes called the oldest molecular rearrangement, [80] but in fact, as we have just seen, the completion of the *process of discovery* did not occur until about 1874–1877, just after the final clarification of the pinacol rearrangement.

6.4.1
Implications of the Rearrangements for Kekulé's Rules of Chemical Change

Thus, Butlerov's demonstration that the rules did not hold in the case of pinacol opened a door into a whole new domain of chemistry. Not only did the resolution of the benzilic acid rearrangement follow quickly but also numerous other rearrangements emerged, notably in the deaminative reactions studied by Victor Meyer and his co-workers. [81]

One can hardly overemphasize the significance of the discoveries of molecular rearrangements. In effect, they showed that in addition to the substitution and degradation reactions considered by Kekulé, one now had to deal with a new kind of transformation, a metamorphosis not by peripheral modification or by molecular fragmentation, but rather by change of the atomic connectivity of the carbon skeleton.

It is not surprising that chemists should feel threatened by these seismic rumblings coming so soon after the upheaval caused by Kekulé's structural theory. The rule of minimal structural change apparently was so deeply ingrained in the consciousness of organic chemists that most of them simply could not accept that it might not be universally true. We can see this in Friedel's reaction when he finally had to admit that a molecular rearrangement had occurred somewhere along the road leading from acetone to pinacol to pinacolone to trimethylacetic acid. He clung to the hope that such rearrangements were anomalies, occurring only under extreme conditions, and that one ordinarily did not have to worry about them. This was just wishful thinking, first because one did not have an independent basis for assigning different reaction conditions to the "extreme" or "mild" categories. In effect, his proposal was circular, since his definition of "mild" conditions amounted to those that did not lead to rearrangement. Second, it gradually became clear that molecular rearrangements occurred under all kinds of conditions, and the study of such processes has become a major branch of chemistry. Kekulé's rules still are used as a guideline for chemical investigations, but nowadays it would be an unwary chemist indeed who failed to keep the possibility of rearrangement in mind.

6.4.2
Why Worry About Proper Attribution of Credit?

If praise for a discovery is inaccurately assigned, the error has a high probability of being propagated in many future articles. The reason for this is that few authors can take the time to make a historical investigation of their own. I think this factor was at work in producing the numerous misassignments of credit in the story of the pinacol rearrangement: At some point, some putatively authoritative reviewer got it wrong, and thereafter, others simply followed. In earlier work, I have committed the sin of insufficient criticality myself in following the crowd by crediting the discovery to Fittig. [82] Actually, the whole chemical literature operates on the assumption that what someone wrote in the past is probably correct, at least until proven otherwise.

Similarly, there has been little if any attempt by later authors to assign credit properly for the benzilic acid reaction. Liebig's role in that case was like that of Fittig in the pinacol rearrangement: Liebig observed the reaction and characterized it empirically (the "that it is"), but it was left to others, decades later, to elucidate the "what it is." To say that Liebig "discovered the benzilic acid rearrangement" or that Fittig "discovered the pinacol rearrangement" implies that they both knew that the reactions they observed were in fact *rearrangements*. As we have seen, neither of them had any such insight. The most that one can say is that they each discovered the *existence* of a *reaction of some kind*, that is, they each fulfilled *half* of the Kuhn criterion.

There is one important difference in the two cases which should now be evident. In the benzilic acid story, no single dominant hero emerged to clarify the whole situation at one brilliant stroke, as Butlerov did in the pinacol case. Instead, a whole group of authors grappled with the benzilic acid chemistry and only gradually, and very reluctantly, eventually reached the answer. One could say that the hero finally was the chemical community itself, whose members perservered (even when such stalwarts as Kekulé retired frustrated from the field) and kept each other honest, without outbreaks of rancor, until the end.

But why is it important to assign credit accurately? After all, there are critics of science who ascribe the scientists' obsession with this subject to our enslavement to the false gods of competition and self-promotion. One response would be that it is good to be first for some easily listed reasons: our grant proposals, our patent applications, our tenure resumés, and above all, our egos. Also, simple justice and fair play demand that we try to make the right attribution. I think all of these motivations are real, although perhaps not all are admirable.

But there is another purpose, one which long has been a concern of historians of science. That is the need to understand the dynamics and temporal course of the process of discovery itself. Only by this kind of analysis is it possible to apprehend the mutual influences of scientific theories and experiments. Moreover, only in this way can the effects of the scientific and cultural milieu of the time on the act of discovery be evaluated. Assigning the wrong credit for a discovery may not only offend against justice, it may distort the historical picture of the event and result in an ignorant, blunted, or superficial description.

In my view, to ascribe the discovery of the benzilic acid rearrangement to Liebig is not justifiable. He had no idea what he had found, nor could he have had at that early time long before the structural theory. Similarly, Fittig, who labored under a disabling adherence to the "conventional" atomic weights of the time, could not have discovered the pinacol rearrangement in its totality.

I think we should attribute the discovery of the pinacol rearrangement to both Fittig and Butlerov. To call Fittig the sole discoverer ignores the crucial change of trajectory of the whole field of chemistry that was going on around that event. It trivializes the great forward leap that chemistry made following the work of Kekulé, Couper, Butlerov, and the other structuralists. At the end of the day, we are obliged to get the assignment of credit right if we wish to understand how science works.

6.5
Appendix. *Ceteris Paribus* or the Problem of Provisos: A Quiet, Menacing Relative of Theory?

6.5.1
Does *Ceteris Paribus* Threaten the Completeness of Science?

The story of molecular rearrangements turns out to be relevant to another issue in philosophy, the concept of *ceteris paribus*. The phrase literally means "all else being equal." [83] *Ceteris paribus* appears to be a widespread if not universal hedge in science, which allows theories to escape refutation under certain circumstances. A generalized form is

> "Most laws are ceteris paribus (cp) laws. If we are being punctilious, what we say is not 'All Fs are G,' but only 'All Fs are G, all else being equal.'" [84]

Despite its expression in Latin, the idea does not seem to have come from classsical natural philosophy. It is only in recent times that a substantial number of philosophers have come to believe that its implications for scientific methodology are profound. [84–90] As is usual in philosophy, others disagree, or agree only with reservations. [91] Nevertheless, the idea lurks on the fringes of science, a surly cousin, not entirely welcome at the dinner table of the philosophy family, potentially hostile, and until recently, seldom acknowledged.

Several authors have pointed out that in most important cases, the hedge under discussion is more specific than is implied by the term "*ceteris paribus* condition." Hempel has proposed that a better terminology is "proviso." The case for keeping the two ideas separate is briefly stated [88] as follows:

> "To say that provisos are just *ceteris paribus* clauses is unhelpful, for the idea of a *ceteris paribus* clause is itself vague and elusive. 'Other things being equal, such and such is the case.' What other things and equal to what? How is the clause to function in theoretical reasoning? Provisos do not call for the equality of certain things, *but for the absence of disturbing factors*." (emphasis added)

In its general form, *ceteris paribus* conditions suffer from the fact that by allowing unspecified exceptions, one risks ending up with a law that has degenerated into a triviality. Thus, "*ceteris paribus* if *A* then *B*" says no more than "if *A* then *B*, or there again, perhaps not." [87] Hempel's point resolves this problem, if it is taken to mean that the proviso must exclude a specific "disturbing factor" that can be independently tested. Although the distinction between provisos and *ceteris paribus* conditions seems clear enough, it is well to keep in mind that many authors don't bother to make it, even though usually what they have in mind corresponds more closely to a proviso.

6.5.2
Provisos, Newton's Laws, and Planetary Astronomy

Provisos become active whenever we interpret a scientific finding. Thus, our interpretation is correct *provided* that no unrecognized factors perturb the observational situation.

To illustrate how such perturbations can obscure the meaning of our results, let me paraphrase a story devised for this purpose by Lakatos. [85] It concerns an imaginary case of "planetary misbehavior." In the period before Einstein's work on relativity, a certain physicist calculates the path of a newly discovered planet by means of Newton's laws of mechanics and gravitation. Astronomical observations show that the actual path deviates significantly from that calculated. However, the physicist does not conclude that the experiment refutes Newton's laws. Instead, he suggests that there must be a previously unknown second planet which perturbs the path of the first. He calculates the mass and orbit of this second planet and then turns to an observational astronomer to find the second planet, but the search is unsuccessful. It seems that the second planet is so small that even the most powerful telescopes available cannot find it. The observational astronomer applies for a research grant to build a still bigger one. Three years later, the bigger telescope is ready. If the unknown second planet were to be discovered now, the event would be greeted with joy as a great victory of Newtonian physics. But alas, even the bigger telescope fails to find the second planet. Does the theoretician now abandon Newton's laws and his idea of the purported second planet? No. He proposes that a cloud of cosmic dust hides the second planet from us. He calculates the location and properties of this purported cloud and asks for a research grant to send up a satellite to look for it. Again, a confirmatory finding would be hailed as a triumph, but the cloud is not found. Does he now abandon Newton's laws? No. He postulates further ingenious auxiliary hypotheses, which are tested in their turn. Eventually, the process peters out in inconclusive speculation, but Newton's laws still are not considered to be refuted. Rather, physicists turn to other problems, leaving the apparent anomalies of the hypothetical planet for future study. Newton's laws are so well accepted that apparent refutations do not suffice to cause their rejection. Note that the physicist in this story is a thoroughgoing Newtonian himself, and all his auxiliary hypotheses are tested within the Newtonian framework. Note also that the hypothetical provisos do not violate Popperian prohibitions against *ad hoc* hypotheses, since they are each testable.

6.5.3
Provisos in the Elucidation of Molecular Structure

Whatever terminology the philosophers use, they seem to agree that the problem of provisos is real in much of science. In chemistry, this is vividly exemplified in the case of molecular rearrangements, as we saw in Chapter 6. There the long delays in the recognition of both the pinacol and benzilic acid rearrangements originated in the investigators' stubborn adherence to Kekulé's rule of minimal structural change. That rule contains an unarticulated, hidden proviso, namely that the rule holds in all cases except those in which it doesn't hold. This is a characteristic feature of all rules in science. When things don't seem to fit in an investigation, a careful search for a hidden proviso in the research protocol may be needed. Finding it not only may explain the anomalies but may often reveal new phenomena that can be tantamount to a discovery.

Similarly, Walter Hückel [92] has clearly pointed out the fallacy of attempts to deduce the detailed bonding structure of benzene from the nature of its reaction products. His admonition is worth quoting at some length. In the following passage, his use of the phrases "excited state" and "excited molecule" do not imply, of course, the common usage of today in which such terms often connote *electronic* excitation. His meaning is unmistakable, however, and the reader will quickly see that he has in mind *chemical* excitation:

> "In all work dealing with investigations as to the bonding states of carbon atoms in aromatic compounds [,] one must not leave out of consideration the changes in bonding states which may occur when chemical reactions attack these atoms [,] the bonding states of which are being studied. By the method of chemical reactions [,] information can be obtained always relative only to the excited state of the molecule. Conclusions as to the state of the bonds in the unexcited molecule cannot be drawn from this information about excited molecules without making a number of rather uncertain assumptions. From the very beginning one must therefore be prepared to find that different kinds of chemical excitation produce different conclusions as to the state of the bonds, so that in some circumstances one is face to face with phenomena bearing a certain resemblance to tautomerism. Furthermore, it is not at all necessary that the physical properties as determined for the unexcited molecule should coincide with those to be expected from the modes of reaction of the excited molecule. For these reasons the otherwise so useful principle of minimum structural change in chemical reactions is to be used only with great care in dealing with problems in which the problem of structure becomes one of the nature of atomic bonding, *i.e.* a problem of constitution."

Another case where an ignored proviso produced a puzzling theoretical result until it was finally recognized is the description of the benzene molecule and other cyclic polyenes in "classical" valence bond (VB) theory, or as it is more widely known, the theory of resonance. [93] For example, this predicted that cyclobutadiene, rather than benzene, should have the highest resonance energy per electron of any of the even cyclic polyenes, in contrast to what was then known of the two species. Further, it gave no explanation for the well-established fact that cyclopentadiene is a much stronger acid than cycloheptatriene. These failures of classical VB theory were among the reasons that Erich Hückel in 1931 turned away from it as a basis for understanding aromaticity and thenceforth adhered to the molecular orbital (MO) approximation, which at that time gave a more satisfactory description. The ignored proviso eventually turned out to be the neglect of cyclic permutions in the truncation of the calculation of energy then used in resonance theory.

6.5.4 Can Theories Be "Complete?"

Some of the concerns are summarized by Cartwright: [86]

> "It is a widespread conviction among scientists as well as philosophers that a fundamental scientific theory should aim at completeness, that is, it should be able to account for everything that happens. ... we have good reason to think that no present nor any yet undreamed of theories are completeable even 'in principle.' The regularities that follow from even the most accurate theories are '*ceteris paribus*' regularities and they hold only so long as all the causes that operate are causes described within the theory. However, ... there are always external causes the theory cannot account for. For these reasons the prospects of completion are rather dim."

Cartwright here expresses pessimism that the age-old goal of a description of nature, in theories that hold without exceptions or conditions, may not be attainable.

Much earlier, Henri Poincaré, [94] had discussed the role of provisos (he used the term *contingencies*) as conditions on the application of natural laws:

> "Then as one can never be certain of not having forgotten some essential condition, it can not be said: If such and such conditions are realized, such and such a phenomenon will occur; it can only be said : If such and such conditions are realized, it is probable that such a phenomenon will occur, very nearly."

Of course, his formulation applies not only if we have "*forgotten*" some essential condition, but also if we were *never aware* of the existence of the condition.

While agreeing that much of science cannot escape the problem of provisos, Earman and Roberts [91] think Cartwright's formulation should not be applied to theories of what they call "fundamental physics." Thus,

> "we argue that the (purported) laws of fundamental physics are not hedged by '*ceteris paribus*' clauses and provisos."

They feel that physicists strive to reach complete theories, and if a particular theory consistently fails to meet that standard, it will be rejected and another one sought.

6.5.5 Fruitful and Fruitless Disregard of Provisos

Although it disregarded a proviso that we now recognize, Pauling and Wheland's version of VB theory during the 1930s enjoyed a period of what Lakatos called "positive heuristic" (see Chapter 3), an era in which the theory continues to make useful suggestions for experiments. In those circumstances, scientists tend to disregard nagging exceptions or known provisos in order to press on with the fruitful work of harvesting the results in more tractable areas. Nevertheless, provisos serve to remind us that there are many improbable but true things in this world. The prudent investigator therefore seeks out the unrecognized provisos in theories and prepares for the unexpected.

6.6 References and Notes

1 A. Kekulé, *Lehrbuch der Organischen Chemie* Ferdinand Enke, Erlangen **1861**. v. 1, p. 1. Rocke (ref. 2) has pointed out that Kekulé's textbook was published first as a series of fascicles, of which the first fascicle of Volume 1 appeared in 1859.
2 A. J. Rocke, *Ann. Sci.* **1985**, *42*, 355.
3 A. Kekulé, *Ann.* **1858**, *106*, 129.
4 A. J. Rocke, *The Quiet Revolution: Hermann Kolbe and the Science of Organic Chemistry*, University of California Press, Berkeley, **1993**.
5 E. S. Wallis, *Organic Chemistry An Advanced Treatise* 1st ed. (Ed.: H. Gilman), Wiley, New York **1938**. v. I p. 723ff.
6 R. Fittig, *Ann.* **1860**, *114*, 54.
7 C. Friedel, *Dictionnaire de Chimie Pure et Appliqué* (Ed.: A. Wurtz), Hachette, Paris **1876**. v. 2, p. 1024.
8 H. E. Roscoe, C. Schorlemmer, *A Treatise on Organic Chemistry* Macmillan, London **1881**.v. III, part 1, p. 632.
9 H. E. Roscoe, C. Schorlemmer, *A Treatise on Organic Chemistry* Macmillan, London **1881**.v. III, part 2, p. 259.
10 A. J. Rocke, *Brit. J. Hist. Sci.* **1981**, *14*, 27.
11 A. Ihde, *The Development of Modern Chemistry*, Dover reprint of the original Harper and Row edition, New York, **1964**.
12 E. Hjelt, *Geschichte der Organischen Chemie von ältester Zeit bis zur Gegenwart*, Vieweg, Braunschweig, **1916**.
13 A. J. Rocke, **1993**: 71ff.
14 A. J. Rocke, **1993**: 136.
15 A. J. Rocke, **1993**: 134ff.
16 W. L. Howard, *Kirk-Othmer Encyclopedia of Chemical Technology*, Wiley, New York, **2000**. On-line version of the 1991 publication
17 J. Liebig, *Ann. Pharm.* **1832**, *1*, 225.
18 Liebig actually carried out three analyses on acetone for C-H. The third, not shown here, agreed with the first two in carbon content, but the hydrogen value he reported cannot be derived from the raw data given in his experimental section. Also, readers who wish to check this conclusion should keep in mind that Liebig used Berzelius atomic weights with C = 12.25.
19 R. Kane, *J. prakt. Chem.* **1838**, *15*, 129.
20 C. Löwig, S. Weidmann, *J. prakt. Chem.* **1840**, *21*, 64.
21 F. Fichter, *Ber.* **1911**, *44*, 1339.
22 R. Fittig, *Ann.* **1858**, *106*, 277.
23 O. Mendius, *Ann.* **1857**, *103*, 39. This is apparently a doctoral thesis carried out under Limpricht's direction.
24 R. Fittig, *Ann.* **1859**, *110*, 23.
25 L. Ebersbach, *Ann.* **1858**, *106*, 262.
26 G. Städeler, *Ann.* **1859**, *111*, 277.
27 A. J. Rocke, **1993**: 172.
28 F. Fichter, *Ber.* **1911**, *44*, 1361.
29 F. Fichter, *Ber.* **1911**, *44*, 1348ff.
30 H. Limpricht, L. v. Uslar, *Ann.* **1857**, *102*, 259.
31 A. Kekulé, *Ann.* **1857**, *104*, 129.
32 A. J. Rocke, **1993**: 168ff.
33 H. Limpricht, *Ann.* **1858**, *105*, 177.
34 R. Fittig, *Ann.* **1861**, *117*, 68.
35 A. Kekulé, *Lehrbuch der Organischen Chemie* Ferdinand Enke, Erlangen **1861**. v. 1, p. 1. Rocke (ref. 2) has pointed out that Kekulé's textbook was published first as a series of fascicles, of which the first fascicle of Volume 1 appeared in 1859.
36 C. Friedel, *Ann.* **1862**, *124*, 324.
37 E. Linnemann, *Ann. Suppl. 3* **1864–5**, 374.
38 C. Friedel, *Ann. Chim. Phys.* **1869**, *[4]* ***16***, 310.
39 C. Friedel, R. Silva, *Ber.* **1873**, *6*, 35.
40 C. Friedel, R. Silva, *Ber.* **1873**, *6*, 70.
41 C. Friedel, R. Silva, *Ber.* **1873**, *6*, 145.
42 C. Friedel, R. Silva, *Ber.* **1873**, *6*, 267.
43 C. Friedel, R. Silva, *Bull Soc. Chim. France* **1873**, *[2] 19*, 98.
44 C. Friedel, R. Silva, *Bull Soc. Chim. France* **1873**, *[2] 19*, 146.
45 C. Friedel, R. Silva, *Bull Soc. Chim. France* **1873**, *[2] 19*, 289.
46 C. Friedel, R. Silva, *Ber.* **1873**, *6*, 826.
47 A. Butlerov, *Justus Liebig's Ann. Chem.* **1873**, *170*, 151(162).
48 A. Butlerov, *Justus Liebig's Ann. Chem.* **1874**, *174*, 125.
49 D. Pawlow, *Ann.* **1879**, *196*, 126.
50 F. Couturier, *Ann. Chim.* **1892**, *[6]26*, 433.
51 L. F. Fieser, M. Fieser, *Advanced Organic Chemistry* Reinhold, New York **1961**. p. 287ff.
52 A. McKenzie, W. S. Dennler, *Ber. Deutsch. Chem. Soc.* **1927**, *60*, 2105.

53 A. McKenzie, R. Roger, W. B. McKay, *J. Chem. Soc.* **1932**, *134*, 2597.
54 K. Mislow, M. Siegel, *J. Am. Chem. Soc.* **1952**, *74*, 1060.
55 C. J. Collins, W. T. Rainey, W. B. Smith, I. A. Kay, *J. Am. Chem. Soc.* **1959**, *81*, 460.
56 W. B. Smith, R. E. Bowman, T. J. Kmet, *J. Am. Chem. Soc.* **1959**, *81*, 997.
57 W. B. Smith, T. J. Kmet, P. S. Rao, *J. Am. Chem. Soc.* **1961**, *83*, 2190.
58 K. Nakamura, Y. Osamura, *J. Phys. Org. Chem.* **1990**, *3*, 737.
59 W. B. Smith, *J. Phys. Org. Chem.* **1999**, *12*, 741.
60 W. B. Smith, *Tetrahedron* **2002**, *58*, 2091.
61 J. Liebig, *Ann.* **1838**, *25*, 1.
62 A. Laurent, *Ann. de Chim.* **1836**, *[2] 61*, 131.
63 N. Zinin, *Ann.* **1839**, *31*, 329.
64 H. Limpricht, H. Schwanert, *Ber.* **1869**, *2*, 134.
65 A. Jena, *Ann.* **1870**, *155*, 104.
66 A. Lachman, *J. Am. Chem. Soc.* **1924**, *46*, 779.
67 E. Grimaux, *Bull. Soc. Chim. Fr.* **1867**, *7*, 369.
68 E. Grimaux, *Ber.* **1869**, *2*, 280.
69 H. Limpricht, H. Schwanert, *Ann.* **1870**, *155*, 59.
70 A. Jena, *Ann.* **1870**, *155*, 77.
71 A. Kekulé, *Ber.* **1870**, *2*, 610.
72 A. Oppenheim, 1869: Footnote to Grimaux's 1869 paper.
73 G. Städeler, *Ann.* **1869**, *151*, 45.
74 T. Zincke, *Ber.* **1871**, *4*, 836.
75 G. Brigel, *Ann.* **1865**, *135*, 171.
76 R. Symons, T. Zincke, *Ann.* **1874**, *171*, 117.
77 A. C. Doyle, *The Sign of Four* John Murray, London **1962**.The Complete Sherlock Holmes Long Stories, Chapter 6. p. 161. Eleventh impression of reprint. Originally published in 1890 by Spencer Blackett, London.
78 A. Baeyer, *Ber.* **1877**, *10*, 123.
79 T. S. Kuhn, *The Structure of Scientific Revolutions* University of Chicago Press, Chicago **1962**. p. 55. Recent historical treatments of the discovery of oxygen are given *inter alia* by Holmes and by Djerassi and Hoffmann: (a) F. L. Holmes, *Antoine Lavoisier – The Next Crucial Year*, Princeton University Press, Princeton, NJ, 1998. (b) C. Djerassi, R. Hoffmann, *Oxygen: A Play in Two Acts*, Wiley-VCH, Weinheim, Germany, 2001.
80 S. Selman, J. F. Eastham, *Quart. Rev. Chem. Soc.* **1960**, *14*, 221.
81 V. Meyer, J. Barbieri, F. Forster, *Ber.* **1877**, *10*, 130.
82 J. A. Berson, *Chemical Creativity: Ideas from the Work of Woodward, Hückel, Meerwein and Others* Wiley-VCH, Weinheim **1999**. p. 117.
83 J. Persky, *J. Econ. Perspectives.* **1990**, *4*, 187. This article contains a brief review of the history of the concept.
84 P. Lipton, *Philosophy* **1999**, *74*, 155.
85 I. Lakatos, *Falsification and the Methodology of Research Programmes*, in *Criticism and the Growth of Knowledge* (Eds.: I. Lakatos, A.Musgrave), Cambridge University Press, New York **1970**. p. 100–103, 186. Reprinted in 1999.
86 N. Cartwright, *Against the Completeability of Science The Proper Ambition of Science* in *London Studies in the History of Philosophy* (Eds.: M. W. F. Stone, J. Wolff), Routledge, London, New York **2000**. v. 2 p. 209ff.
87 P. Mott, *Mind* **1992**, *101*, 335.
88 C. G. Hempel, *Provisos: A Problem Concerning the Inferential Function of Scientific Theories*, in *The Limitations of Deductivism* (Eds.: A. Grünbaum, W. C. Salmon), University of California Press, Berkeley, CA **1988**. p. 19–36.
89 J. A. Fodor, *Mind* **1991**, *100*, 19.
90 R. N. Giere, *Laws, Theories, and Generalizations, in The Limitations of Deductivism* (Eds.: A. Grünbaum, W. C. Salmon), University of California Press, Berkeley, CA **1988**. p. 37–46.
91 J. Earman, J. Roberts, *Synthese* **1999**, *118*, 439.
92 W. Hückel, *Theoretical Principles of Organic Chemistry*, Elsevier, Amsterdam, The Netherlands **1955**.v. 1, p. 645. Translated by F.H. Rathmann from the corrected 7th German edition.
93 J. A. Berson, *Chemical Creativity: Ideas from the Work of Woodward, Hückel, Meerwein and Others* Wiley-VCH, Weinheim **1999**. p. 44–50. A review.
94 J. H. Poincaré, *The Value of Science The Foundations of Science* G. B. Halsted, translator, The Science Press, Lancaster, PA **1913**. p. 355. Reprinted in 1946.

7
Some Non-Refutative Motivations in Science.

7.1
Science Without Refutation

Whatever one's opinion of Popper's methodological system, that it is incomplete seems beyond dispute. Many advances in science do not originate in attempts to falsify a theory, nor can they be used after the fact for such a purpose.

Lakatos [1] describes the structure of an ideal falsificationist test as requiring

> "a two-cornered fight between theory and experiment so that in the final confrontation only these two face each other ..." and "the only interesting outcome of such confrontation is (conclusive) falsification ..."

Then he quotes another formulation of falsificationism by Joseph Agassi, at one time a staunch adherent of that doctrine:

> " [The only genuine] discoveries are refutations of scientific hypotheses." [2]

Lakatos goes on to his own contrary view that

> "... history of science suggests that tests are -—at least – three-cornered fights between rival theories and experiment," and "some of the most interesting experiments result, *prima facie*, in confirmation rather than falsification."

Any working chemist will know that the explorers of great tracts of the chemical research landscape populated their territory without falsificationist motivations. I mention briefly here only two such vast regions. The first is the common but important activity of empirical data-gathering and correlation; the second is the enterprise of chemical synthesis.

7.2
Correlation: The Search for Regularity in Nature

Gower [3] summarizes some motivations for scientific research unencumbered by any obligation to falsify a theory as follows:

> "... experiments are not always tests of theories or hypotheses. Evaluating an innovative technique, exploring a novel topic, measuring or re-measuring a physical constant, identifying the characteristics of a new synthetic chemical, or simply observing the the effects on animal behavior of controllable variables, are all legitimate modes of experiment. Sometimes, indeed, experiments are designed simply so that we can get nature to behave in unfamiliar ways and thereby illuminate its more familiar ways. We have, in short, reasons for "intervening" in the world, or "interrogating" nature, other than that of finding out whether our guesses are correct or not."

One characteristic that many of these activities have in common is that they operate without any real theory at all. (I neglect here the trivial argument that the theory being tested in these data-collecting experiments is just that some order or regularity will be observed). One recognizes this as essentially a *pre-theoretic* phase of an inductivist investigation in which correlations of experimental facts into some sort of relationship form the necessary observational foundation for the *subsequent proposal* of a theory.

Thus investigators try to find the basis to allow the proposal of theories that would eventually explain the trends they discovered in the laboratory. Popper considered this kind of activity less significant than refutation, an evaluation chemists always have ignored. The list of examples is long, and it includes many empirically derived correlations that have proven to be of great significance. These include studies of structurally characteristic IR frequencies (e.g., CO and OH groups), [4] NMR chemical shifts and coupling constants, [5] rules for correlation of UV-Vis absorption wavelengths with molecular structure, [6] etc. It also includes virtually all botanical and animal taxonomic classificatory correlations with the molecular structures of naturally occurring chemical constituents such as terpenes, alkaloids, and pheremones. In none of these cases was there a theory that guided the search until much data had been accumulated.

7.3 Accidental and Serendipitous Discoveries

In the course of such searches (but not only there), one sometimes finds an unexpected result or a discovery that requires an unanticipated confluence of circumstances, often one that appears only under special conditions that could not have been specified in advance. These may be called "accidental," or when they are found while seeking some other disparate objective, "serendipitous," discoveries. [7] Chapter 6 described two such accidental discoveries: Liebig's unwitting choice of conditions for the reaction of benzil with alkali which led to the rearrangement to benzilic acid, a set of circumstances that escaped the notice of Laurent and prevented him from making the discovery first, and Fittig's attempt to alkylate acetone which led to the dimeric reduction product pinacol. We now briefly describe two more famous, – one might say canonical — examples: Pasteur's resolution of the enantiomers of sodium ammonium tartrate, and Wöhler's synthesis of urea from ammonium cyanate. Both of these have been well documented by others, so that our discussion here will emphasize only the accidental qualties of each.

7.3.1
Louis Pasteur and Chirality [8]

The creative investigator often is the one who welcomes these surprises, recognizes that, through no conscious human action, a new question has been put on the table, and then devises appropriate ways to find the answer. Pasteur, who during his brief career in the field, was one of the greatest organic chemists (and soon thereafter was to become the embodiment of chemistry's greatest sacrifice to medicine), said truly "Chance favors the prepared mind." I venture to suggest that the "preparation" Pasteur was thinking of was precisely the knowledge that the unexpected could emerge without warning. An example is his most famous chemical experiment, the separation of the two mirror image forms of sodium ammonium tartrate. This achievement, the first deliberate resolution of enantiomers – in fact, the first demonstration of the existence of enantiomers – depended on the curious, rare, and then unprecedented fact that, under exactly the right conditions (sharply limited and *fortuitously* discovered), the two forms separated from solution as mirror image crystals. Others later found that this crystallization property for this particular salt emerged only at laboratory temperatures below 27 °C. It just happened that Pasteur carried out his experiments at a time when the Parisian climate matched this requirement. His dramatic follow-up experiments that led to the idea of molecular chirality form one of the most inspiring events of 19th-century chemistry. One moral of the story is, as any French schoolchild knows, "*Aide-toi, le ciel t'aidera,*" ("Heaven helps those who help themselves.")

7.3.2
Wöhler and the Synthesis of Urea from Ammonium Cyanate [9]

In the early part of the 19th century, chemists put much effort into the preparation of new compounds. An important motivation was to explore the relationship between chemical composition (the number and kind of atoms in a molecule) and chemical properties. In 1828, the young Friedrich Wöhler, following this path, reported [10] an attempt to prepare prepare ammonium cyanate (**1**) by mixing ammonium chloride **2** with silver cyanate **3** by a standard "double decomposition" or exchange reaction. The reaction instead gave silver chloride **4** and a crystalline product which he subsequently identified as urea **5** (Eq. 1).

NH_4CNO (**1**)
NH_4Cl (**2**) + $AgCNO$ (**3**) $\rightarrow$ $AgCl$(**4**) + H_4N_2CO (**5**) (1)

What theory was Wöhler testing when he synthesized urea? When that discovery was made, the structural theory of organic chemistry would not be put forward for another 30 years. One could anticipate what reaction to expect by comparison with other precedent observations of what were thought to be similar combinations, and the reaction product was characterized simply by physical properties (crystalline form, melting point, etc.) and atomic composition. For the sake of alignment with Popper's formulation, we might call this the theory of chemical analogy. Thus the metathesis ("double decomposition" or exchange) reaction Wöhler tried to achieve, (eq. 2)

$$NH_4Cl\ (\mathbf{2}) + AgCNO\ (\mathbf{3}) \rightarrow AgCl(\mathbf{4}) + NH_4CNO\ (\mathbf{1})\quad (2)$$

was analogous to other such known metatheses, for example, ammonium sulfate (**6**) plus barium chloride (**7**) to give ammonium chloride (**2**) plus barium sulfate (**8**) (eq. 3).

$$(NH_4)_2SO_4\ (\mathbf{6}) + BaCl_2\ (\mathbf{7}) \rightarrow 2NH_4Cl\ (\mathbf{2}) + BaSO_4\ (\mathbf{8})\quad (3)$$

One could say that, formally, when Wöhler found urea as his reaction product instead of the one he expected, ammonium cyanate, the theory of chemical analogy was refuted, even though that was not Wöhler's objective.

However, it seems to me far more important that Wöhler's discovery brought forth three different and entirely unexpected results. First, as the first total synthesis of a natural product, it was a forerunner of the eventual (but not immediate) demise of the theory of vitalism in organic chemistry, that is, the idea that the organic compounds of biological origin, such as urea, could be formed only through the intervention of a "vital force." Second, Wöhler noted that ammonium cyanate (**1**) and urea (**5**) have the same number and kind of atoms, yet ammonium cyanate should have been a salt and should have liberated ammonia instantly when treated with alkali, whereas urea did not have these properties. In other words, these must be what we now call *isomers,* different compounds with the same atomic composition but with entirely distinct properties. This was the first recognized example of this phenomenon. Thirty years later, with the establishment of the structural theory of organic chemistry, the difference between the two was recognized as a difference in *connectivity* of the atoms: ammonium cyanate is $NH_4C{=}N{=}O$, and urea is $NH_2\text{-}CO\text{-}NH_2$. This provides the third unexpected result from Wöhler's experiment, namely the transformation of one isomer into the other is an example, probably also the first, of a *molecular rearrangement.*

We must remember that the structural theory taught *what strucures* could exist, but said little about *what reactions* could occur. Thus, even after the structural theory had appeared, many new and unexpected reactions were discovered, often as a result of a failure of analogy. Again, there are too many examples to enumerate here, but one thinks immediately of the discovery (Chapter 6) of the formation of pinacolone in the attempted acid-catalyzed dehydration of pinacol (a carbon skeletal rearrangement), Cannizzaro's discovery of the disproportionation of benzaldehyde to benzyl alcohol and benzoate ion in alkaline solution, and in fact, a great number of the standard synthetic reactions of organic chemistry, many of which were discovered in the pre-mechanistic era.

In the Wöhler discoveries flowing from the ammonium cyanate-to-urea reaction, no theory under test there was relevant to the new insights revealed by his experiments. In the era when the nature of molecular structure did not yet exist – in fact, at a time when some influential chemists rejected the very thought that one could ever know the relative positions of the atoms in a molecule – the concept of a molecular rearrangement could not exist, and therefore there could be no theory to explain it. Wöhler's experiment revealed the need for a whole new kind of thinking about molecules. As to whether the experiment was considered "accidental" or surprising at the time, we need only note what Wöhler recorded: [10]

> "... this investigation has given the unexpected result (*'unerwartete Resultat'*) that the combination of cyanic acid with ammonia leads to urea, a remarkable fact, noteworthy in that it shows an example of the synthetic generation of an organic, and actually so-called animal, compound from inorganic compounds."

7.4 Are Accidental Discoveries Rare?

With this background, we can examine another of Popper's corollaries, which may be conveniently identified as the "few accidental discoveries" rule.

Popper's position on this issue may be summarized in his own words:

> "Accidental discoveries occur too, of course, but they are comparatively rare". [11]
>
> "... even most so-called 'chance-discoveries' are as a rule refutations of theories which were consciously or unconsciously held: they are made when some of our expectations (based upon these theories) are unexpectedly disappointed. Thus, the catalytic property of mercury was discovered when it was accidentally found that in its presence a chemical reaction had been speeded up which had not been expected to be influenced by mercury. But neither Oersted's [effect of an electrical current on a magnetic needle] nor Röntgen's [X-rays] nor Becquerel's [spontaneous radioactivity] nor Fleming's [penicillin] discoveries was really accidental, even though they had accidental components: every one of these men was searching for an effect of the kind he found." [12] (I have shown in brackets the subjects of the discoveries to which Popper presumably referred).

Skeptical readers already will have noted that Popper simply brushes aside the single example he listed of an accidental discovery, as though this was an exception to the overwhelmingly more numerous and significant cases of discoveries from searches for effects of the kind actually found.

Elsewhere [13] Popper expands his hypothesis:

> "Even the careful and sober testing of our ideas by experience is in its turn inspired by ideas; experiment is planned action in which every step is guided by theory. We do not stumble upon our experiences, nor do we let them flow over us like a stream. Rather, we have to be active: we have to *'make'* our experiences. It is we who always formulate the questions to be put to nature; it is we who try again and again to put these questions so as to elicit a clear-cut 'yes' or 'no' (for nature does not give an answer unless pressed for it). And in the end, it is again we who give the answer; it is ourselves who, after severe scrutiny, decide upon the answer to the question which we put to nature ..."

This passage seems to attribute to the experimenter a high degree of conscious awareness of the question being asked. Perhaps Popper is saying that this is the way one *should* do science, but if it his intention to portray the way in which science *does happen*, I would have to say that he is very far away from real life in the laboratory, and this distance distorts his view of the scientific experience. He stresses that we ask the questions, and we induce Nature to give us the answers. At least as often, however, it is the other way around: Nature gives us the answer to a question we have not consciously asked, and it is only eventually that we realize what the question was.

A quotation from Feyerabend is relevant here: [14]

> "... the actual development of institutions, ideas, practices, and so on, often *does not start from a problem* but rather from some extraneous activity, such as playing, which, as a side effect, leads to developments which later on can be interpreted as solutions to *unrealized* problems." (emphasis added).

This point of view gives a very different picture of the actual conduct of science from that proposed by Popper.

One reason for Popper's insistence on the unimportance of accidental discovery may be that he sees it as part of his fight against inductivism. Thus, what both Wöhler and Pasteur were aiming at was a sort of accretion of instances, Wöhler to determine the trends in properties as a function of the molecular composition, and Pasteur to link crystal form with properties, both essentially inductivist actions.

One might argue that the whole issue is just a matter of semantics, that is, that Popper is defining the problem out of existence by the stipulation that whenever we carry out an experiment we are always testing some theory, whether or not we are conscious of it. One form of this might be that the "theory" is merely the conviction that some governing principle or at least a regularity of behavior in the field of inquiry actually exists, even though we don't know what it is. This rationalization will not stand scrutiny. To propose that Popper was simply playing word games with us trivializes his thinking. Whatever one may think of his philosophy, Popper was not a trivial man.

I offer the suggestion that Popper found the idea of accidental discovery profoundly disruptive to his epistemological scheme. If the major activity of scientists is to be the testing of theory by refutations, and if this is to be the most effective way that science can move forward, as I believe Popper's writings claim, then other kinds of discovery must be of only secondary importance. Anything that does not fit this scheme has to be portrayed as mundane or of lesser significance. If one believes in such an epistemology, I suppose it is not impossible simply to ignore the numerous counterexamples or to provide rationalizations that make them disappear. My experience as a working scientist convinces me that, on the contrary, accidental discovery is an abundant well-spring of new ideas in science. Without its continuing nurture, a science limited to "conjectures and refutations" would become a cold, arid intellectual wasteland.

7.5
Organic Synthesis, a Corroborative Enterprise

Synthesis is another area in which one cannot say that the motivation is to refute a theory. For the present purpose, we can focus on two branches of synthesis, both of which involve the preparation of molecules in a series of deliberate, experimenter-effected steps. These are the synthesis of molecules to test the limits of structural possibility and the synthesis of naturally occurring molecules of complex structure. Each of these steps must be carefully planned and executed in order to reach the target molecule. [15]

7.5.1
Synthesis of Molecules to Test Theories of Chemical Bonding and Reaction Mechanism and the Synthesis of Natural Products

Ever since the rise of the structural theory of organic chemistry in the period around 1858–1860 (see Chapters 4 and 6 and references therein) and the theory of stereochemistry around 1874, tests of their validity and range of applicability have furnished an inexhaustible source of motivation for chemists. In most instances, the logical structure of the synthetic efforts that resulted was not refutative but rather exploratory.

As chemists find themselves stirred by the mysterious allure of the symmetrical and the beautiful, aesthetic and self-challenging motivations also become apparent in many such instances. Like the impulse driving the heroic geographical expeditions, the urge to explore often is mixed with a sheer will to surmount risk and hazard in order to triumph over adversity. The parallel that comes to mind is George Mallory's famous answer when he was asked why he wanted to climb Mount Everest: "Because it is there."

I choose just a few of many beautiful examples (Scheme 1): [16] For the purpose of the discussion to follow, let me put these into two different categories. In one category, there is Willstätter's monumental synthesis of cyclooctatetraene **9** (Scheme 1), [17;18] in which the scientific motivation was a test of the then reigning theory of aromaticity, but in which we also can glimpse the challenge in the difficult completion of a multistep synthetic plan. As just three examples in the second category, I would place the syntheses of the small ring hydrocarbons [3]prismane **10**, [19], cubane **11** [20], and (1.1.1)propellane **12.** [21]

9

10

11

12

Scheme 7.1

In an earlier era, synthesis traditionally held an honorable place in chemistry as a valid *confirmation* of a proposed molecular structure. We saw two instances in Butlerov's work, described in Chapter 6. It is true that in the last few years, the ability to determine the structure of a complex molecule to a high degree of probability has occurred at a much faster pace than the ability to synthesize it, which has somewhat diminished the prominence of that traditional role of synthesis. But synthesis continues to serve another purpose, namely in the preparation of molecules of unusual or unknown structure.

Let's examine for a moment the objectives of such a synthesis. In the case of cyclooctatetraene **9**, the question was will this molecule be like benzene, then the only other known conjugated cyclopolyene? Benzene was unique for its "aromatic" properties (reaction by substitution rather than addition, special stability, etc.). Or would cyclooctatraene behave like an ordinary olefin?

In the case of the small-ring compounds **10–12**, the question was can they even exist? The enormous strain built into these molecules by the contraction of the bond angles away from the normal tetrahedral angle of 109? 26' poses a daunting synthetic challenge. In compound **12**, for example, each bridgehead carbon projects all four of its bonds to one side of a plane, rather than in the normal directions toward the corners of a tetrahedron. In fact, when [3]prismane **10** eventually was synthesized (fortunately only in small quantities), the answer to the question of whether it exists was found to be "Just barely," as the molecule itself rudely but convincingly announced: When ground, crystals of the substance decomposed with explosive brissance!

The reader must now be asking: What is the significance of the two categories I proposed earlier, namely cyclooctatetraene **9** as opposed to the small-ring hydrocarbons **10–12**? It is that the purposes of the syntheses now can be seen to have different relationships to the formal structure of falsificationist principles. The cyclooctatetraene case can be fitted to Lakatos's idea (Section 7.1) of a "three-cornered fight between rival theories and experiment." Here the rival theories are that cyclooctatetraene is either aromatic or non-aromatic. Experiment shows that it is non-aromatic, and hence the rival theory is defeated, or falsified.

However, no such fit exists in the case of the small-ring compounds **10–12** or in many other synthetic endeavors. The purpose of their synthesis cannot be cast into falsificationist format. In each of those examples, one of the main questions was: Is this compound capable of existence? If one tries to make this conform to conjecture and refutation, one might say that one conjecture is that the compound can exist. A successful synthesis would be a confirmation of the conjecture. However, an unsuccessful synthesis, a failure to synthesize, cannot be taken as a refutation. The failure may be caused not by the compound's inability to exist or spontaneous instability, but rather by the chemist's inability to devise or carry out a suitable synthetic design. A failure to synthesize thus embodies no information relevant to falsificationist concerns.

It should be obvious that this incongruence of the activity of chemists with the method of conjectures and refutations is wide-spread in the field of chemical synthesis. In many instances, chemists engaged in synthesis are simply not trying to falsify any hypothesis. On the contrary, they are trying to *verify* something. What are the things they are trying to verify?

We can examine this in the context of the motivation for synthesis of large organic natural products. Nowadays this often is claimed to be the preparation of larger quantities of rare physiologically active materials. However, many syntheses of even a moderately complex natural product would be totally impractical for that purpose and never are used that way. Instead, often the motive is to demonstrate the authors' command of organic chemical knowledge in developing new synthetic steps or strategies.

It is sometimes said that a total synthesis is a stringent test of theory. This may be so, provided that one is willing to accept *verification* as a test. The theory being tested is, implicitly, the one that says that organic reactions take place according to certain well established rules of chemical combination and reaction. Thus, at every stage of a synthesis, the chemist pits personal knowledge of these rules, and his or her judgment of their range of applicability, against the task of making the next step work.

These would not be tests of any theory by Popper's criterion, since "every genuine test of a theory is an attempt to falsify it, to refute it." On the contrary, the chemist's most fervent wish is that theory will be *verified*. The whole synthesis depends upon an unbroken string of such verifications. Usually, the result will depend on whether the designed steps are *chemospecific,* that is, compatible with the portion of the synthetic molecule that, one hopes, will not become inadvertently entangled in the synthetic steps.

Just because of the ramified and often multifunctional structure of the reactants in such syntheses, this is a question that usually cannot be answered definitively in advance, so there is no alternative to trying the step. If unintended side reactions occur, all one will have shown is that the hoped-for compatibility was not there. Thus, the outcome of the attempted synthesis does not constitute a falsificationist test of theory. Even if the synthesis fails at some point, this does not suffice to refute the theory of regularity of chemical behavior. It simply means that the chemist did not have the knowledge to foresee the possibility of a side reaction lurking in a complicated molecular array. A failed synthesis therefore is neutral with respect to the issue of refutation.

If I am the synthetic chemist in this example, the only theory that I will have tested is my *hunch,* based on experience and knowledge, that the step would work. I suppose it could be said in jest that in that case the theory being tested is "all Yale professors of organic chemistry are smart enough to have made the correct prediction," but one may doubt that that is what Popper had in mind when he wrote of scientific theories.

One might argue that the synthesis of molecules of potential medicinal importance is simply an adjunct to a test of a theoretical prediction that such and such a type of structure will have desirable pharmacological properties. The test of theory, in this view, then does not really start until the synthesis of the required test compounds has been achieved. However, the pharmacological test itself certainly is aimed at corroboration, not refutation. Pharmaceutical chemists do not spend much of their time devising refutations of a theory that predicts therapeutic activity for a class of compounds. The whole purpose of the enterprise is to find active compounds, so they are looking for confirmation, not refutation.

7.6 Reflections

It seems clear that much of the activity of synthesis seems to fall outside of the doctrine of conjectures and refutations. That set of guidelines envisions no important role for the largely confirmative aspect of an activity that has proven central to the concerns of chemists. Whether one wants to make a certain compound to confirm its structure, or to make a medicinally significant molecule, or to test a theoretical existence issue, or just because "it is there," the point remains that the actual synthesis itself frequently has an inescapably confirmative purpose, not a refutative one.

In trying to adapt falsificationist doctrine to the field of chemistry, we seem to have uncovered something arrestingly curious. If, as Popper says, confirmations count only if they are intended to test a theory, and if, again as he says, a valid test of theory must be refutative, we are led to a dilemma: if a confirmative synthesis succeeds, it doesn't count, but as any chemist knows, if it fails, it doesn't refute.

Synthesis is probably the most active area of chemical science and has been for more than a century. But, as a corollary of the method of conjectures and refutations, we have just derived the strange result that much of synthesis has no justifiable claim to call itself a major participant in the advancement of chemical knowledge. I cannot imagine that many chemists will accept this conclusion.

The arguments given in this chapter exemplify the *indispensability* of accidental discoveries and of verificationist procedures for the healthy progress of chemistry. I daresay that they also apply in other disciplines. We do science for all kinds of motivations. Empirical data-gathering expeditions, whose results usually are not designed or interpreted *ex post facto* as tests of theory, and chemical syntheses with a variety of objectives, have always been and will continue to be close to the center of chemistry. Chemists cannot be expected to give up the fruitful joys and challenges of those modes of research in order to conform to a falsificationist agenda

7.7 References and Notes

1 I. Lakatos, *Criticism and the Growth of Knowledge* (Eds.: I. Lakatos, A. Musgrave), Cambridge University Press, Cambridge, U.K. **1999**. p. 115. Reprint of the 1974 revised edition, originally published in 1970.

2 J. Agassi, *Impulse* **1959**,*3*, 2. As cited by Lakatos, ref. 1.

3 B. Gower, *Scientific Method: An Historical and Philosophical Introduction* Routledge, London **1997**. p. 242.

4 L. J. Bellamy, *The Infrared Spectra of Complex Molecules* Wiley, London **1954**. p. 205ff.

5 N. S. Bhacca, L. F. Johnson, J. N. Shoolery, *NMR Spectra Catalog*, Varian Associates, **1962**.

6 L. F. Fieser, M. Fieser, *Steroids* Reinhold, New York, NY **1959**. p. 15ff. A review.

7 R. M. Roberts, *Serendipity*, Serendipity: Accidental Discoveries in Science,Wiley, New York, **1989**.

8 J. D. Bernal, *Science and Industry in the Nineteenth Century*, Routledge and Kegan Paul, London, **1953**; G. L. Geison, *The Private Science of Louis Pasteur*, Princeton University Press, Priceton, NJ, **1955**; A.W. Hofmann, *Ber.* **1882**, *15, 3148*;

G. B. Kaufmann, R. D. Myers, *The Chemical Educator* **1998**, *3*, 1430; G. B. Kauffman, I. Bernal, H.-W. Schütt, *Enantiomer* 1999, *4*, 33; A. Ihde, *The Development of Modern Chemistry* Harper and Row, New York **1964**. p. 321ff. Dover reprint of the third printing of 1970.

9 A. Ihde, *The Development of Modern Chemistry* Harper and Row, New York **1964**. p. 170ff. Dover reprint of the third printing of 1970; G. Bugge, *Das Buch der Grossen Chemiker* Verlag Chemie, Weinheim **1965**. V. 2, p. 31ff.

10 F. Wöhler, *Ann. Phys. Chem.* **1828**,*88*, 253.

11 K. R. Popper, *The Logic of Scientific Discovery* K. R. Popper, J. Freed, L. Freed, translators, Routledge, London **1959**. p. 108. From *Logik der Forschung,* the original German ediition of 1934 (reprinted 1992).

12 K. R. Popper, *Conjectures and Refutations: The Growth of Scientific Knowledge* Routledge and Kegan Paul, London **1963**. p. 220. Reprinted by Routledge in 1996 from the revised fifth edition of 1989.

13 K. R. Popper, **1959**: p. 280.

14 P. Feyerabend, *Against Method: An Outline of an Anarchistic Theory of Knowledge* New Laft Books, New York **1975**. p. 154. Reprinted in 1997 by Verso as *Against Method* from the third edition of 1993.

15 Note that our discussion here leaves out of consideration another important branch of synthesis, namely the preparation of polymeric materials. These typically are the products of a repeated occurrence of individual steps, that is, a chain, triggered by an appropriate stimulus, in which the individual steps are not under the direct control of the experimenter but result from the chemical properties of the component monomers, the nature of the initial stimulus, and the conditions of the reaction.

16 H. Hopf, *Classics in Hydrocarbon Chemistry: Syntheses, Concepts, Perspectives,* Wiley-VCH, Weinheim, Germany, **2000**. This book gives an excellent survey of syntheses of unusual hydrocarbons.

17 R. Willstätter, E. Waser, *Ber.* **1911**,*44,* 3423.

18 R. Willstätter, M. Heidelberger, *Ber.* **1913**,*46,* 517.

19 T. J. Katz, N. Acton, *J. Am. Chem. Soc.* **1973**,*95,* 2738.

20 P. E. Eaton, T.W. Cole, Jr., *J. Am. Chem. Soc.* **1964**,*86,* 962.

21 K. B. Wiberg, F. H. Walker, *J. Am. Chem. Soc.* **1982**,*104,* 5239.

8
False but Nevertheless Fruitful Theories. Speculations on the Biogenesis of Alkaloids

Pierre Duhem's definition of a false theory (see Chapter 2) serves to introduce this chapter:

> "A false theory is not an attempt at an explanation based on assumptions contrary to reality; it is a group of propositions which do not agree with the experimental laws. *Agreement with experiment is the sole criterion of truth for a physical theory.*"

There is a wide-spread assumption that false theories must be rejected. One notable formulation of this comes from Richard Feynman: [1]

> "In general we look for a new law by the following process. First we guess it. Then we compute the consequences of the guess to see what would be implied if this law that we guessed is right. Then we compare the result of the computation to nature, with experiment or experience, compare it directly with observation to see if it works. If it disagrees with experiment it is wrong. In that simple statement is the key to science."

This is a reduced and over-simplified ("naïve") version of formal Popperian falsificationism (see Chapter 3), but many of the scientists who call themselves followers of Popper probably have nothing more sophisticated than Feynman's rule in mind when they do so. It is true that Popper himself [2] realized that such a procedure, which seems to advocate a death sentence for a theory when it has failed a test of *only some* of the statements derived from it, could not be imposed. Rather, he conceded that the abandonment of a theory relies upon "the scientific instinct of the investigator." However, although all of the statements *derived* from a theory may be known, further statements may be *derivable*. Beyond that, a theory may fail a test not because it is entirely erroneous, but because it is valid only under a certain range of circumstances. Chapter 5 of this book already has described such a case, and further cases are not hard to find. [3] Consequently, new and unforeseen consequences and predictions may emerge, and it is clear that this injection of a degree of sophistication into the "naïve" version of falsificationism still leaves us with no guidance on just when or how the final process of falsification is to take place. The problem seems to be that it is not always easy to tell when a theory is false, that is, the "scientific instinct of the investigator" may be unreliable.

Nevertheless, I sense that even among those who do not consciously subscribe to falsificationism, the rejection of false theories is accepted as axiomatic. The idea seems self-evident: Who cannot see that persistence in false belief will lead one onto the path of error? As Thomas Kuhn and others have emphasized, however, the rejection of a false theory, especially of one that has been widely influential, usually takes place over a period of time and becomes effective only by consensus.

Events in science rarely can be reduced to simple syllogisms. The struggles of scientists to understand a group of phenomena occur in real time. Theories, experiments, doubts, refutations, and confirmations usually follow each other in disorderly, jumbled rout. It is a curious but significant fact that during this phase, certain theories, even some of those subsequently rejected on indisputable grounds, distinguish themselves by their utility. They bring a measure of order or predictability to events. Whether they ultimately survive is a separate issue.

We now examine some cases in which false theories led to useful and even startling valid discoveries. For example, the related case in Chapter 5 mentioned above was Meerwein's theory of the 6,2-shift mechanism in bicyclic terpene rearangements. In that instance, the original experimental basis for the theory soon was overthrown, and hence, by formal criteria, the theory should have been rejected. Nevertheless, many cases exemplifying the 6,2-shift eventually came to light. Thus, the theory was correct, even though the original reason for proposing it was not.

The present chapter, however, concerns a fundamentally different type of theory, a theory that is truly false, as it were. Thus, in using the term "false," I do not refer to theories like Meerwein's, whose domain of applicability was narrower than he imagined, nor like Kekulé's original benzene structure, which required an *ad hoc* subsidiary hypothesis to make it coincide with known facts (see Chapter 4), nor like the approximations of the early organic quantum chemists, which were refined by later advances. Rather, for the present purpose, I shall define a "false" theory as one that is demonstrably incorrect in some or all of its most fundamental concepts. An instructive example is the so-called "Woodward fission" theory of the biosynthetic origin of a large class of plant bases derived from tryptophan. Important examples include the indole-derived alkaloids, such as the strychnos, cinchona, and iboga groups, and the ipecac alkaloids. Woodward's theory was an ingenious modification of earlier theories of the biogenesis of alkaloids, among which the foremost was that of Robert Robinson.

8.1
Early Theories of Alkaloid Biogenesis

The alkaloids have long been objects of special fascination for chemists because of their physiological action, and later, because of the abundant variety of complex chemical structures represented among them. Late in the 19th century, chemists began to solve these structural problems analytically, and in some cases, actually to carry out laboratory syntheses to confirm the solutions. They found that, with the methods of the time, such syntheses usually were demandingly difficult, which led them to ask:

How does Nature go about synthesizing these intricate molecules in the living plant with such apparent ease? Further questions soon emerged: What physiological role do the alkaloids play in the life of plants? What insights can be gained into the intricacies of plant metabolism and into biological processes generally from an understanding of the biosynthesis of alkaloids and other natural products?.

The deep exploration of these puzzles at the biological level ultimately required the application of sophisticated biochemical and genetic tools. However, these were not available until long after the challenges had been recognized, and the contributions of biochemistry to the field of the biogenesis of alkaloids did not begin to be decisive until after World War II, when both stable and radioactive isotopes of several elements, as well as the instruments to make quantitative measurements of isotopic abundances, became widely available. Instead, the early advances in biosynthetic investigation came from the organic chemists. Their researches were driven by their knowledge of chemical reactions, combined with an anthropocentric and not always explicitly articulated conviction that Nature probably made use of similar reactions in biosynthesis. In other words, they made conjectures on the actual path of *in vivo* synthesis on the basis of analogy to their own synthetic experience in the laboratory. These early efforts were starting points for the more comprehensive theories of Robinson in 1917 (see Section 8.1.4). Since Robinson acknowledged this stimulus only cursorily and without details or references, we briefly outline some of the early history here.

8.1.1
Alkaloids from Proteins: Pictet's Theory

One of the first such speculations [4] was that of the Swiss chemist Amé Pictet (1857–1937). [5;6] In the period 1905–1906, he proposed that alkaloids were formed from the products of breakdown of proteins, *i.e.*, amino acids, in the plant. These products, he suggested, would combine with other endogenous substances to form alkaloids, which would be deposited in the plant tissues because the plant has no excretory organs.

Pictet [4] noted the widespread occurrence among the plant alkaloids (*e.g.*, nicotine **1**, ricinine **2** (Scheme 8.1) of the pyridine nucleus, a six-membered nitrogen-containing ring, and conceived the idea that an important coupling partner would be formaldehyde, which had been detected as an actual constituent in green plants. In his hypothesis, formaldehyde, when combined with amino acids from protein breakdown, could lead to the pyridine ring.

OCH_3 CN N O CH_3 N N CH_3

1 nicotine

2 ricinine

Scheme 8.1

About a decade later, Pictet and Chou [7] tested this idea *in vitro* by heating the protein casein in aqueous hydrochloric acid for 6 hours while adding formaldehyde dimethyl acetal. They obtained (as might have been expected!) a complicated reaction mixture, from which they indeed were able to isolate pyridine **3**, 2,6-dimethylpyridine **4**, isoquinoline **5**, 4-methylisoquinoline **6**, and several other incompletely identified bases (Scheme 8.2). They called attention to the fact that although isoquinolines were among the reaction products, *quinoline* **7** or its substituted derivatives were not. In a control experiment, they showed that none of the heterocyclic bases were formed if the formaldehyde dimethyl acetal was omitted from the reaction mixture. Decades later, isotopic tracer studies [8; 9] showed that the biogenesis of pyridine rings in nature was much more complicated than Pictet had imagined.

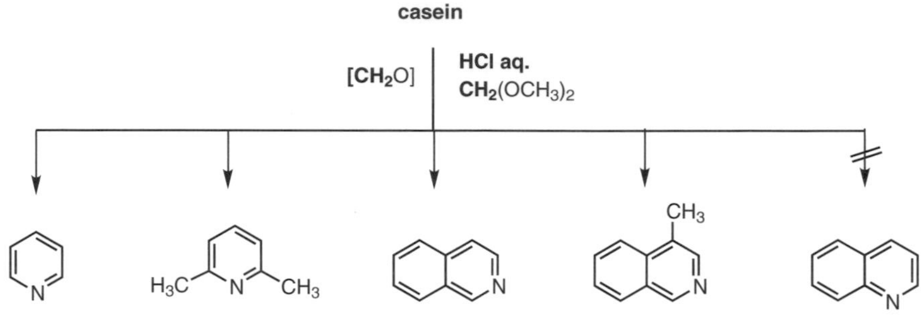

3 **4** **5** **6** **7**

Scheme 8.2

8.1.2
The Mannich Reaction, a Key to Synthesis and Biosynthesis

A few years later, in 1911, Pictet was to return to the problem of isoquinoline rings in alkaloids [10]. Among the important results of that work was the Pictet-Spengler synthesis of tetrahydroisoquinolines **10, 11** from formaldehyde (or other aldehydes) and β–arylethylamines **8, 9** (Scheme 8.3).

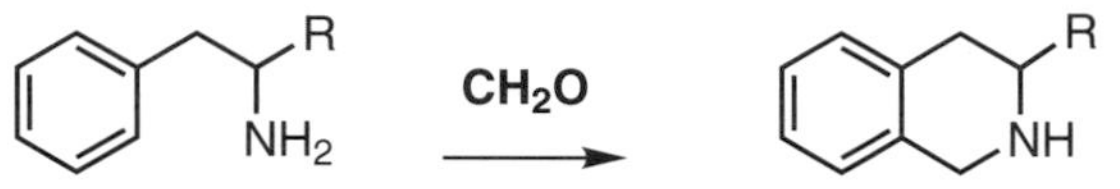

8: R = H
9: R = CO_2H

10: R = H
11: R = CO_2H

Scheme 8.3

With the advantage of hindsight, we recognize that this is an example of what came to be known as the Mannich reaction [11;12], that is the condensation of an amine, an aldehyde, and a nucleophilic center (an aromatic ring or an enolic site, as in –COCHR- or O_2NCH_2-). In fact, the first examples with such ketones had been reported by Tollens and co-workers [13] in the period 1903–1906 (as in the formation of **12** from acetophenone, ammonia, and formaldehyde, Scheme 8.4), and independently by others during 1906–1908. [11;12] A number of authors had observed the reaction with nitro compounds even earlier [11]. The designation "Mannich reaction" presumably honors that author's later (1917) [14] recognition of its generality and his development of it for a broad range of synthetic application.

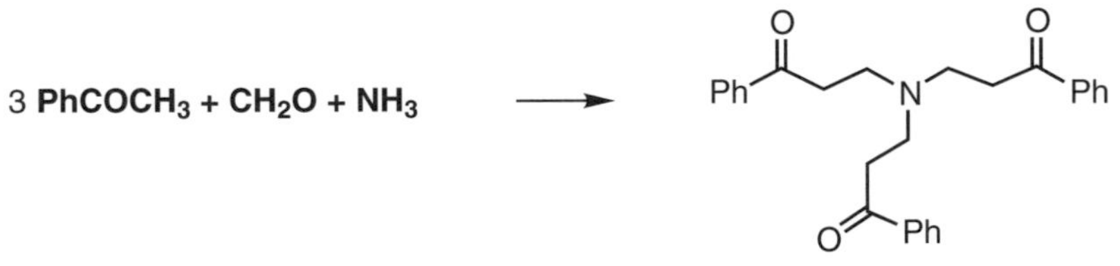

$$3\ PhCOCH_3 + CH_2O + NH_3 \longrightarrow$$

12

Scheme 8.4

In 1912, Mannich was a professor in the pharmaceutical institute of the University of Berlin. His first experiments [15] in the condensation of α-carbonyl C-H bonds with formaldehyde and ammonia came as the result of the chance observation (his word was *"Zufall"*) that when the widely used analgesic antipyrine (**13**, Scheme 8.5) was mixed in aqueous solution with hexamethylenetetramine (a source of formaldehyde and ammonia), a crystalline insoluble precipitate formed almost immediately. The molecular formula of this material showed that it had been formed by the condensation of three molecules each of formaldehyde and antipyrine and one molecule of ammonia, with elimination of three molecules of water. At this time (1912), Mannich and his co-worker Krösche apparently were not aware of the similar and earlier observations by Tollens and others mentioned above.

Hydrolysis of the product gave back the original reactants formaldehyde, ammonia, and antipyrine. Mannich and Krösche clearly recognized that the hydrolysis products and the condensation products were reversibly interconvertible. They considered two structures for the condensation product (**15** and **16**, Scheme 8.5). In structure **15**, each antipyrine ring is linked to a methylene group through a bond at C_4, whereas in **16**, the bond is through oxygen. They favored the O-methylene-bonded structure **16** largely because they thought it improbable that the carbon-bonded structure could be hydrolyzed to the observed products, formaldehyde, ammonia, and antipyrine:

> "Presumably, the bond between the methylene group and the carbon atom of antipyrine could not be hydrolyzed at all, and that between the methylene group and the nitrogen only with difficulty."

This was a self-contradictory conclusion, since whatever the mechanism, the existence of a path for the condensation by dehydration must imply the existence of a reverse path for the hydrolysis. Certainly, important examples of hydrolysis or alcoholysis of carbon-carbon bonds already were well-known in the reverse aldol and acetoacetic ester condensations. Moreover, the oxygen-linked structure **16**, which they

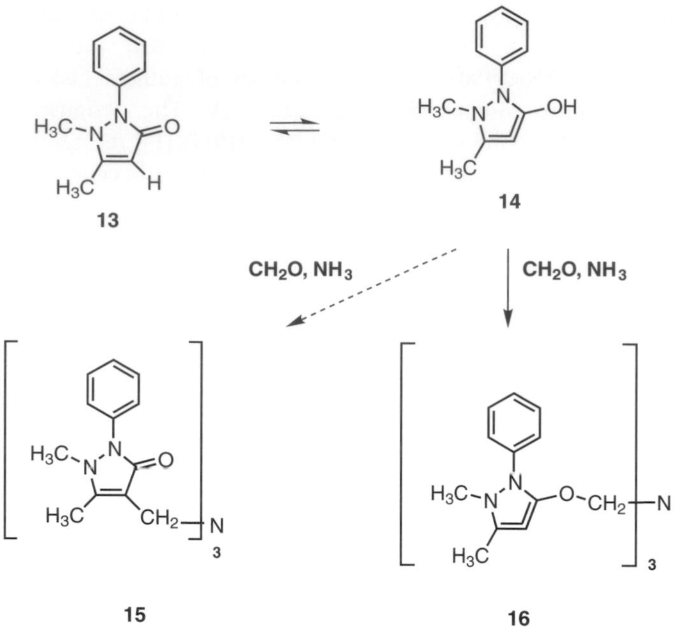

Scheme 8.5

proposed to be formed by way of the enol **14**, we recognize today as unacceptably strained, since incorporation of the allene unit into the five-membered ring requires a contraction of the bond angle from 180⁰ to near 108⁰.

Mannich's 1917 paper [14] was delayed by the war. By that time, he had become aware of the 1905 paper of Schaefer and Tollens, [13] which had assigned the structure of a typical "Mannich" product from acetophenone as **12**, shown in Scheme 8.4, containing C-methylene bonds rather than O-methylene bonds. Mannich adopted this idea immediately, switched his assignment of the antipyrine product (Scheme 8.5) to the C-methylene structure **15**, and in further work consistently assigned such structures to his amine-formaldehyde-α-carbonyl condensation products.

8.1.3 Winterstein and Trier

In 1910, the year prior to Pictet's tetrahydroisoquinoline synthesis, Ernst Heinrich Winterstein and Georg Trier, chemists at the Agricultural-Chemical Laboratory of the Swiss Federal Institute of Technology (E.T.H.), published a book, *Die Alkaloide,* [16] in which they proposed what Birch [17], writing decades later, called the "surprisingly modern" idea that a number of alkaloids may be formed in nature from the condensation of a phenylethylamine with an aldehyde. A specific suggestion put forward by Winterstein and Trier was that the then hypothetical reaction of a dimethoxylated phenylalanine **17** with homoveratrylacetaldehyde **18** could give rise to the tetrahydroisoquinoline product **19** by what we now would call an intramolecular Mannich reaction(Scheme 8.6).

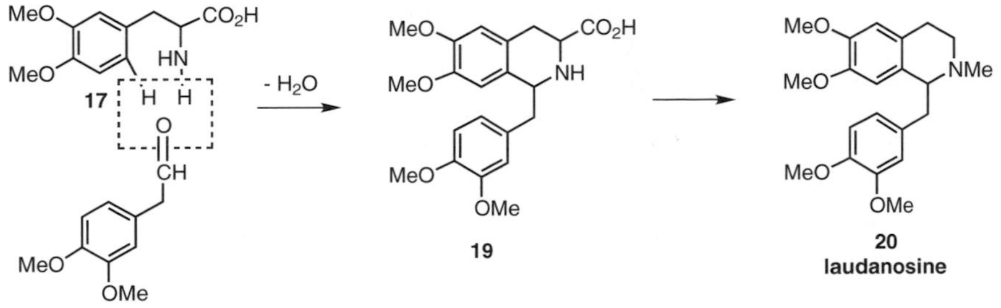

Scheme 8.6

The significance of the dashed box shown in the scheme, which appears in the original paper and was intended to indicate the source of the molecule of water formed in the reaction, will be discussed further below (see Section 8.3) Upon decarboxylation and N-methylation, **19** would form the known alkaloid laudanosine **20**, a minor component of the basic fraction from opium and hence botanically related to morphine.

Birch [17] states that Winterstein and Trier were somewhat vague as to what functionality would be "chemically necessary" for their proposed building blocks to be assembled in the plant, although he admits that they recognized the need for and even in some cases the nature of the appropriate functionality. He ascribes to Robinson and Lapworth the credit for the explicit formulation of chemically plausible pathways. Birch attributes this lack of specificity in the Winterstein-Trier treatment to the supposed fact that they were "plant physiologists and not...chemists." Perhaps this issue reduces to a matter of definition, but it should be noted that Winterstein's entry in the *Historisch-Biographisches Lexicon der Schweiz,* 1934 edition, lists his professional metier as "chemiker" (chemist) and his professional activities as "author of numerous articles in the field of plant-agricultural- and physiological chemistry." Winterstein (1865–1949) was a professor at E.T.H., and Trier was an assistant there. To me, the Winterstein-Trier chapter on biogenesis gives assurance of a professional level of competence in organic chemistry. It is true that their suggestion of lysine as the biogenetic precursor of the pyrrolidine (hygrine and cuscohygrine) and tropane (atropine) alkaloids was chemically implausible, whereas Robinson's identification of ornithine as the likely precursor has stood the test of later biochemical studies (see Section 8.8.1). Of course, one would have to agree that Winterstein and Trier were less sophisticated chemists than Robinson, but there were few chemists of the time to whom that uncomplimentary comparison would not apply.

Winterstein and Trier were fully aware of Pictet's theories and actually commended him for "interesting and stimulating thoughts and experiments" in the field of alkaloid biogenesis [16]. The major disagreement between them concerned the nature of the aldehyde component. Winterstein and Trier postulated an oxygenated phenylacetaldehyde (or alternatively, an oxygenated phenylpyruvic acid) in the biogenesis of the isoquinoline alkaloids (see Scheme 8.6). Pictet preferred to use formaldehyde instead, because it had been identified in living systems, whereas phenylacetaldehydes had not. Pictet's scheme of course would require that the resulting 1-unsubstituted

Figure 8.1 Robert Robinson.
© The Nobel Foundation.

tetrahydroisoquinoline be further transformed to reach the benzyl tetrahydroisoquinoline structure of laudanosine and other related alkaloids.

8.2
Robinson: The Mannich Reaction, Tropinone, Biomimetic Synthesis, and Biogenesis of Alkaloids

In 1917, at the age of thirty-one, Robert Robinson (1886–1975) published two brilliant papers [18;19] which made him the leading figure in the biogenetically motivated structural analysis of alkaloids. Robinson (shown later in life in Figure 8.1) was then at the University of Liverpool, but Birch [17] makes a strong case for the proposition that the seminal ideas and even some of the experimental work came during the period 1913–1915 at the University of Sydney. Some of the ideas owed much to Robinson's collaboration in the early stages with Arthur Lapworth.

Another root of Robinson's thinking may be traced back even a few years before his sojourn at Sydney, to a period when he was associated with William Henry Perkin, Jr. (1860–1929), then the professor of organic chemistry at Manchester (see Section 8.3).

The first of the two papers [18] was entitled *A Synthesis of Tropinone;* the second [19] was *A Theory of the Mechanism of the Phytochemical Synthesis of certain Alkaloids.* Birch argues, [17] on what seem to be plausible grounds, that although the intellectual connection between the two is often presented as if the theoretical paper motivated a test of its validity in the synthetic paper, the sequence of Robinson's thoughts may well have been the reverse. That is, the original motivation for the synthesis was different, and it was only after the synthesis had been achieved that its relevance to the theoretical development became apparent to him. This is clearly seen in the statement from the *second* 1917 paper [19]:

"The synthesis of tropinone recently described [18] on account of its simplicity, is probably the method employed by the plant."

Scheme 8.7

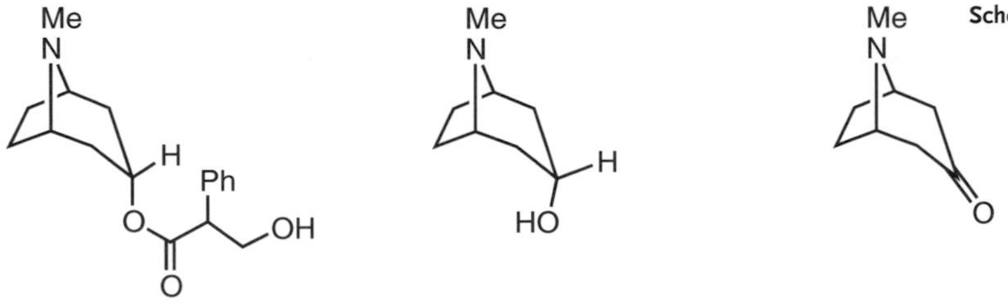

21 **atropine** **22** **tropine** **23** **tropinone**

Among the several solanaceous plants that produce medically useful alkaloids, the most important for our present story is the deadly nightshade, belladonna (*Atropa belladonna*), which gives rise to the ester atropine **21** (Scheme 8.7), the racemic form of another alkaloid, hyoscyamine. In commerce, atropine is formed from hyoscyamine by racemization during isolation. The alcohol of atropine is tropine **22**, which can be oxidized to the ketone tropinone **23** and regenerated from it by reduction. Thus, tropinone was a key substance in the elucidation of the structure of these bases, which had been worked out, primarily by Willstätter and by Merling in the early 1900s, [9;20;21] and potentially it was also a tempting synthetic target for the preparation of artificial atropine-related drugs.

By 1903, Willstätter had accomplished the first total synthesis of tropinone **23**. [22] This was a monumental achievement at the time and was the culmination of many years of work in this field, but the actual synthetic route was laborious in the extreme. It started with suberone (cycloheptanone), itself not readily available in quantity at that time, and required 17 steps to reach tropine, whose interconversion with tropinone, as we have said, was already well known. No further total syntheses of tropinone appeared for more than a decade after Willstätter's.

Imagine then the impact made upon the field in 1917 by the appearance of Robinson's paper [18] reporting a dramatic total synthesis of tropinone (Scheme 8.8).

His procedure afforded that ketobase in a single step from readily available starting materials! Robinson visualized this as a Mannich reaction (without specific mention of it), a "one-pot" preparation in which the addition of succindialdehyde **24** and methylamine **25** gave the double carbinolamine **26**. This then reacted *in situ*, without isolation of the product, with the calcium salt of acetonedicarboxylic acid **27** to give the bicyclic dibasic salt **28**. Acidification led to decarboxylation and formation of the ketobase tropinone **23**. (Today, following Lapworth, we probably would formulate the detailed intermediate steps in a slightly different way, with iminium ion intermediates **29** and **30**).

Within ten days of the receipt of Robinson's paper on the tropinone synthesis, the *Journal* received a second paper [19] from him entitled *A Theory of the Phytochemical Synthesis of certain Alkaloids*. In this second paper, Robinson essentially adopts the

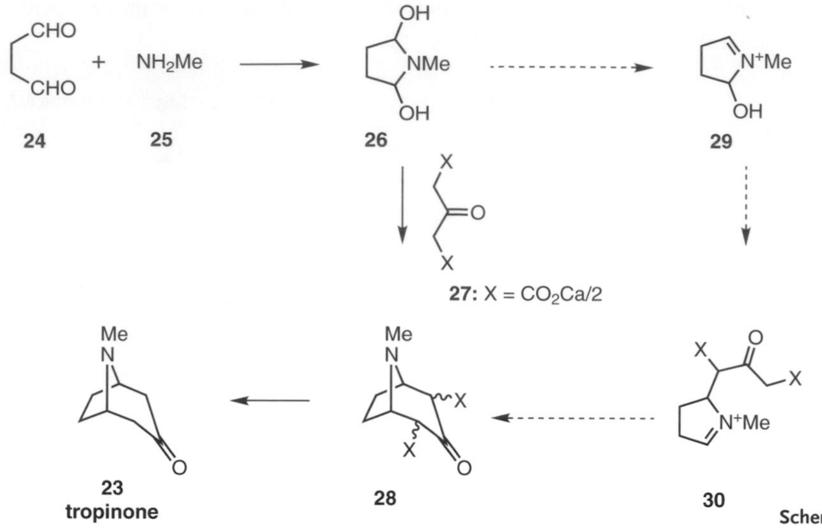

24 25 26 29 27: X = $CO_2Ca/2$ 23 tropinone 28 30

Scheme 8.8

proposals of Pictet and of Winterstein and Trier, namely that the amino acids are the likely biological precursors of the alkaloids. Curiously, his acknowledgment of their priority is crabbed and ungenerous. Thus, Robinson disposes of their efforts by referring to them only obliquely in his introduction:

> "Although in recent years, largely owing to the investigations of Pictet and his collaborators, there has been a due recognition of the importance of the role played by formaldehyde in the production of alkaloids in plants, and although it is generally admitted (compare Winterstein and Trier, *Die Alkaloide*, pp. 263–317) that the amino acids and carbohydrates are the most probable starting points for the majority of phytochemical syntheses, yet little progress has been made in ascertaining the nature of these processes or even in the less ambitious task of formulating possible mechanisms based on laboratory analogies."

This is surely an obscure way of of describing the history. Winterstein and Trier's hypothesis for the formation of laudanosine by a Mannich reaction (Scheme 8.6) was the prototype for much of the Robinson scheme. Further, the experimental studies leading to the Pictet-Spengler synthesis of the tetrahydroisoquinoline nucleus (Scheme 8.3) clearly involved Mannich chemistry very similar to Robinson's own laboratory synthesis of tropinone and to his proposals for the biosynthesis of that alkaloid and many others. By the time of his Pedler Lecture to the Chemical Society in 1936, Robinson [23;24] was ready to make a more open acknowledgment to Winterstein and Trier, but Pictet now had dropped out of the picture entirely and his name does not appear in the historical review.

With a bit of magnanimity, one can perhaps overlook Robinson's neglect of the usual rules of scientific deportment. Two circumstances might be considered exculpatory. First, with regard to the omission of Pictet's contribution in Robinson's 1936 review, it may be that Robinson, like most of us in similar circumstances, was unable to resist the opportunity for a bit of acccount-settling. The course of the controversy is given in some detail in Robinson's autobiography, [25] but a brief summary will suffice to make the point here. Robinson, during the period around 1909, when he was still a student in Perkin's laboratory, had achieved a synthesis of the alkaloid papaverine **34** from **31** and **32** by the route shown in Scheme 8.9.

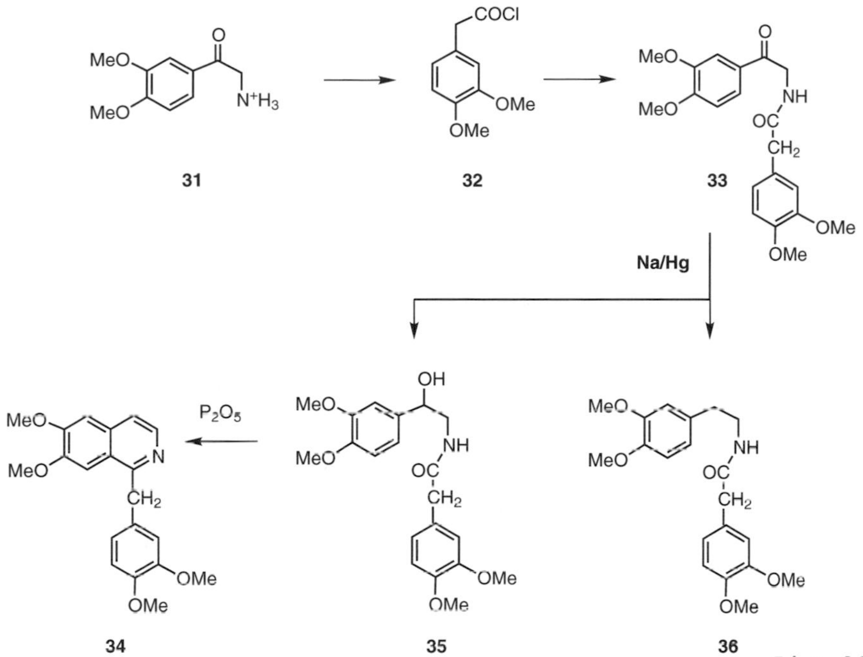

Scheme 8.9

The reduction of the ketoamide **33** gave an oily material, which upon treatment with phosphorus pentoxide gave the natural base **34**. Robinson was eager to publish this result, but Perkin restrained him, because the hydroxyamide **35** had not been isolated in pure form from the oily intermediate. While Robinson struggled to complete this phase, Pictet and Gams [26] reported a total synthesis of papaverine by essentially the same route, thereby anticipating Robinson and Perkin. Although he accepted the validity of the Pictet-Gams synthesis, Robinson regained a bit of ground in the contest when he showed that the crystalline material, which Pictet thought to be the hydroxyamide **35** and had isolated from the reduction of the ketoamide **33**, was not **35** but instead was the deoxygenated derivative **36** formed as a side product. Robinson concluded that Pictet's report that he had converted this crystalline material to papaverine by the method of Scheme 8.9 could not be correct, and that Pictet and Gams actually had used the oily intermediate containing a mixture of **35** and **36** in the final step. Robinson also eventually isolated the alcohol **35** in crystalline form. Pictet challenged the accuracy of Robinson's studies, but it soon became clear that Robinson was correct on these points. Perhaps these experiences might have left Robinson less than kindly disposed toward Pictet.

A second reason for Robinson's occasional deficiency of attention to the work of others may originate not in personal grudges but rather from his impatience with the whole enterprise of tracing the history of ideas, even his own. Birch explains Robinson's attitude:

> "I once asked him about the evolution of his classical ideas on the electronic theory of reactions: [His answer was] 'I cannot remember, but what does it matter? The result is what counts.'"

Someone who thus considers the origins of his own ideas to be unworthy of much thought might have the same opinion of the thoughts of others.

We can see this disregard in Robinson's 1917 paper on biogenesis of the alkaloids. [19] He makes the point that in his proposed biogenetic schemes, carbon-carbon bond formation takes place by only two processes: the aldol condensation and "the very similar condensation of carbinolamines, resulting form combination of an aldehyde or ketone and ammonia or an amine, and containing the group C(OH)N, with substances containing the group CHCO." This is clearly the Mannich reaction, but Robinson makes no reference to Mannich in that paper, despite Mannich's extensive publications starting as early as 1912. At first sight, this it is difficult to suppress a reaction of disapproval at what seems to be a usurpation. Soon, however, one realizes that Perkin and Robinson carried out even earlier extensive studies beginning in 1910 on just such reactions of carbinolamines, especially with cotarnine **37**, a degradation product of the opium alkaloid narcotine **40**. In fact, these authors had achieved a laboratory total synthesis (Scheme 8.10) of narcotine by a Mannich-like condensation. [27–30]

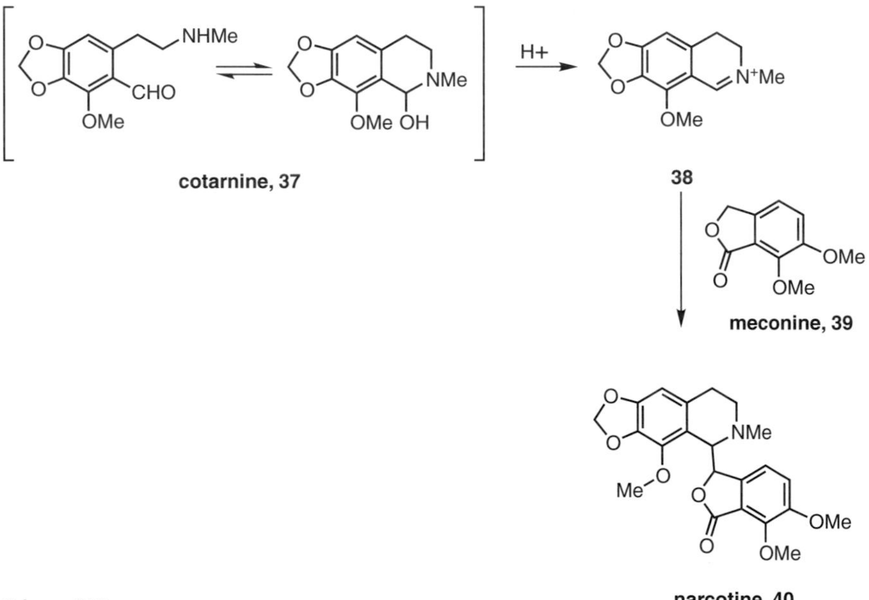

Scheme 8.10

Robinson [19] mentions these studies, but without a literature reference and in such an oblique way as to befuddle the reader. Part of the communication gap here comes from Robinson's unfortunate and life-long predilection for the passive voice in referring to his own work, perhaps motivated by a misguided effort to avoid appearing boastful. This makes it difficult for the uninitiated reader to determine who actually was responsible for a particular observation. In any case, the point is that Robinson was far from scrupulous in reference not only to the work of others but also to his own.

Thus, in the period 1910–1912, the early development of the Mannich reaction had many contributors. It is a curious fact that none of the protagonists (Perkin and Robinson, Winterstein and Trier, Mannich, and Pictet and Spengler) mentions the relationship of their own work in the field to that of any of the others. We might consider three possible reasons for this: (1) reluctance to relinquish credit for priority; (2) the occurrence of one of those periods in science when an idea is "in the air", and several discoverers seize upon it; (3) failure to recognize the relevance of other work. Given human nature, point (1) is difficult to dismiss, not only here, but generally. Point (2), although frequently invoked in the literature of history of science, is merely a restatement of the problem, unless the historian is able to provide reasons why that particular idea should have emerged at that particular time. I think that point (3), however, provides a useful basis for conjecture, and in fact, may contain implications for point (2) also.

8.3 Reaction Mechanisms in the Early 20th Century

During the period that the Mannich reaction was emerging to prominence as an important pathway in biosynthetic speculation, mechanistic reasoning was almost completely undeveloped. The closest that most organic chemists came to it was in their use of dashed boxes (see Scheme 8.6) to indicate which atoms were being "split out" in a condensation reaction. Modern readers will recognize that this was really no more than a device to keep track of the reaction stoichiometry and overall structural result. The idea of a reactive intermediate had been applied in only a few scattered cases, [31] and the primitive theoretical framework obscured mechanistic similarities between reactions, similarities that might lie hidden under superficial differences in structure or reaction course.

We can glimpse the thought processes of the time in a few sentences by Robinson [32] about his mentor, William Henry Perkin, Jr. (1860–1929):

> "He had a simple outlook on the subject, which he treated as a system in which the parts were related by reactions. However, he never paused to consider the underlying mechanism of organic chemical reactions. For him, chemistry was still the science for transformation of materials and these could be represented by changes in structural formulae. Synthetic and other reactions could usually be explained by drawing rings around reactive moieties, with consequent changes in the binding of bonds."

For example, we can follow the thinking that ultimately led in 1911 to the Mannich-related Pictet-Spengler synthesis of tetrahydroisoquinolines (Section 8.2). It is quite clear that in Pictet's mind, this really was an extension of his work of 1909 [33;34] on the synthesis of dihydroisoquinolines, which in turn was a methodological modification of the Bischler-Napieralski reaction, a synthesis of dihydroisoquinolines by dehydration of a β–phenylethylamide(*e.g.*, **41** → **42**, Scheme 8.11). Pictet and Finkelstein [34] then achieved a synthesis of the tetrahydroisoquinoline alkaloid laudanosine **20** by reduction of the dihydroisoquinoline intermediate **42** (Scheme 8.11).

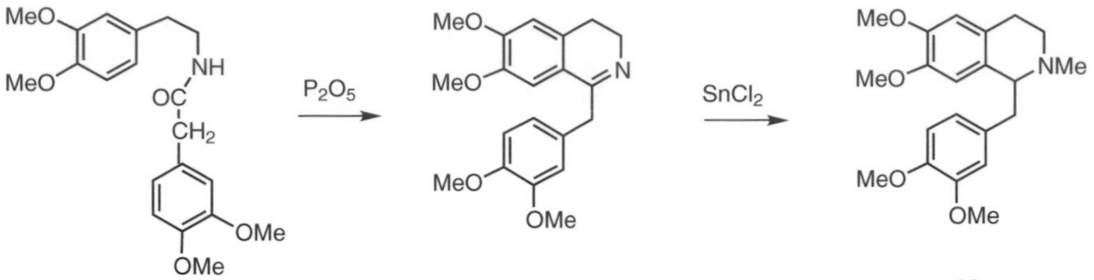

MeO, MeO, NH, OC, CH_2, OMe, OMe — **41** — P_2O_5 → MeO, MeO, N, MeO, OMe — **42** — $SnCl_2$ → MeO, MeO, NMe, MeO, OMe — **20** **laudanosine**

Scheme 8.11

The 1911 Pictet-Spengler synthesis of tetrahydroisoquinolines (Scheme 8.3) looks very similar to a Bischler-Napieralski reaction in the structural sense and differs only in the oxidation state of the carbonyl component. It would not be surprising if, at that moment, this resemblance seemed stronger than any resemblance to the Mannich condensations.

8.4
From "the Mechanism of the Phytochemical Synthesis of Certain Alkaloids" to "Structural Relations of Plant Products"

Robinson's initial convictions on the biogenesis of alkaloids were intuitive. The tropinone synthesis had been achieved in aqueous solution using reagents plausibly derivable from cell constituents, near room temperature, and at nearly neutral pH, a set of circumstances that later would be referred to as "physiological conditions." As we have seen, Robinson accounted himself impressed by the "simplicity" of the reaction, which he implied was necessary for *in vivo* synthesis.

At first, he suggested that these schemes served not only to highlight structural relations between putative reactants such as the amino acids and the final product alkaloids, but also truly identified the pathway of biosynthesis in the living plant. This phase lasted from his first paper in 1917 until well into the 1950s. But by his own account, [35] events soon after that changed his mind. Thus, regarding the biogenesis of alkaloids, he states

> "In general, the idea of my work on structural relations was to dissect the molecules so as to show the relation with simpler natural products and, if possible, to use in the dissection the reverse of the reaction which might conceivably occur

> [in nature]. As the whole of this work was speculative in nature I can only look back on it with the feeling that it may have given a stimulus to later workers, for at least they had something in black and white which they could disprove, and very often they did just that. However, quite a considerable amount of the original remains, unless one takes the view that the schemes put forward were actual biogenetic syntheses. Although the general idea of the specifications of natural products according to their structural relation is undoubtedly sound, the speculative basis of the operation had to be replaced by experimental methods."

Robinson makes it clear that the change in his thinking came after work by Battersby and a number of others had led to "most surprising conclusions." We discuss these in Section 8.10.1. In my opinion, Robinson's characterization of the role played by his speculations as generally fruitful is surely sound, even though many of them were incorrect in detail and a few entirely off the mark.

8.5 The Strychnine Story

8.5.1 Structure

With this background, we are ready to examine the most famous of the wrong theories that grew out of his speculations, namely the idea that the key reaction in the biogenesis of the non-indolic portion of the indole alkaloids incorporates an oxygenated phenylalanine residue. What Robinson referred to as "most surprising conclusions" was the eventual demonstration that this idea is entirely false. Nevertheless, I intend to make the case that even this theory had its fruitful consequences.

Strychnine **49** (Scheme 8.12), a poisonous alkaloid responsible for the medicinal action of *nux vomica,* is undoubtedly the most thoroughly investigated of the plant bases. For decades, the molecule resisted attempts to determine its diabolically intricate structural formula. Among the names of the numerous chemists who struggled with this problem we find those of luminaries such as William Henry Perkin, Jr., Hermann Leuchs, Robert Robinson, Heinrich Wieland, Vladimir Prelog, and Robert Burns Woodward, the latter four of whom eventually became Nobel Laureates.

The work progressed in stages, until in 1931, K.N. Menon and Robinson tentatively proposed structure **43**, which was the first attempt to incorporate all of the known properties in a complete formula. Details of the work leading to this conclusion have been summarized by Holmes, an associate of Robinson. [36;37] Menon and Robinson were aware that the parts of the formula were not uniformly well established, and indeed, in the following year, they were obliged to take into account new evidence that the dihydroindole nucleus of strychnine could not contain the grouping N-CH-CH. Their revised formula **44** accommodated this requirement.

However, this formula too was soon called into question by the report of Leuchs that the oxidation product diketonucidine (**45** on the basis of **44** as strychnine) could

43
Menon and Robinson, 1931

44
Menon and Robinson, 1932
Holmes and Robinson, 1939

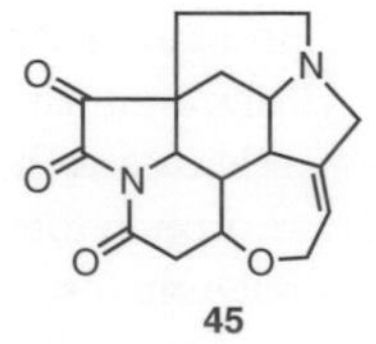

45

46
Blount and Robinson, 1932

47
Robinson, 1947

48
cinchonine

49
Woodward, Brehm, and Nelson, 1947
Chakravarti and Robinson, 1947

Scheme 8.12

be brominated readily, presumably in the position α to a carbonyl group. Blount and Robinson therefore, with some reservations, suggested the formula **46** for strychnine. By 1939, other evidence cast serious doubt on **46.** Especially persuasive was the work of Holmes and Robinson, [38] who repeated the Leuchs bromination experiment and showed that the procedure actually gave different products but not the α-brominated derivative. Accordingly, Holmes and Robinson reverted to the earlier formula **44,** which remained the favored expression for several years.

8.5.2
A Digression on Structural Elucidation

Before going on to the conclusion of the story of the constitution of strychnine, I briefly consider the standards of structural elucidation in organic chemistry. There are critics who have chided Robinson for making so many disposable attempts to reach a final answer. This position implies a set of rules of the game analogous to those which a student taking an examination must follow. Once the answers have been handed in for marking, no changes are allowed. Thus, nature is the stern teacher who punishes wrong answers with a bad grade.

There is no doubt that by consensus of the community, particular opprobrium attaches to the publication of a structure in the literature that turns out to be incorrect.

I believe this is because such an error is usually highly visible, and therefore the blunderer may feel intense embarrassment. The reason for this is that, unlike many other mistakes, which may be arguable in such a way that the author's failure can be mitigated, a wrong structure is just plain wrong, and everyone knows that to be the case. The structural framework of organic chemistry is so rigid that it leaves no room for approximations.

This attitude is evident in the impact of Wagner's discovery of the terpene rearrangements [39] on the famous German chemist Adolf von Baeyer. Baeyer had spent many years in efforts to elucidate the structure of these molecules and had assigned a number of structures to important members of the series unaware of the frequent occurrence of the Wagner rearrangements. When Wagner's work appeared, Baeyer published a paper in which he ruefully referred to it in the words "according to the new theory, nearly all of the structural formulas I proposed are incorrect." Baeyer's student Willstätter gives a special reason for Baeyer's chagrin in this sketch: [40]

> "Baeyer never anxiously avoided that which so many of us feel is unscientific, namely the proposal of a hypothesis that is insufficiently supported by experimental evidence and that is certain to have only a short life. He employed hypotheses for their utility. However, structural formulas he held to be extremely important and even definitive things, the content of which was indeed for the most part independent of the theories of our time and therefore was fixed, even when profoundly radical developments caused us to express our views in a new format."

However, in my view, it is unrealistic to demand that chemists be judged, in all circumstances, by such a harsh and unforgiving standard. Of course, today, when such powerful techniques as multi-dimensional nuclear magnetic resonance spectroscopy, high resolution mass spectrometry, and single crystal X-ray analysis are routinely available for the accurate solution of the structures of most molecules of reasonable size, the publication of an incorrect structure often is associated with a failure to apply properly the tools that lie to hand. In those circumstances, a presumption of negligence or ignorance may well be an appropriate judgment.

On the other hand, Robinson and the other strychninists worked in an era when the tools were very limited. The most important quantitative analytical method was the determination of elemental composition by combustion. Even the simplest spectroscopic methods were late entrants in the development of the strychnine structure. To deal with a problem as difficult as strychnine, investigators had to invent new experimental and interpretive techniques at every turn. I think that Robinson published his tentative formulas in the full knowledge that they were just working hypotheses. They served to summarize concisely the current state of the investigation, to focus the thinking of his laboratory, and to inform the chemical community of progress. It would have been impractical for him to have withheld periodic reviews in favor of waiting until he was firmly convinced that there would be no further revelations demanding a revision.

8.5.3
The Endgame of the Strychnine Structural Problem

The structure **44** (Scheme 8.12), to which Robinson had reverted in 1939, was not destined to survive. In Britain, work on strychnine and other problems was seriously disrupted by World War II, and it was not until 1946 that the next results began to appear. [41] On January 14, 1947, Robinson submitted a cryptic letter to Nature, which appeared in the February 22 issue. [42] In it, he proposed the new and startling formula **47** (Scheme 8.12) for strychnine. Note that the lactam ring is now five-membered instead of six-membered as in all the previous formulas. Moreover, the dihydroindole ring bears hydrogens at both C-2 and C-3, thus contradicting the previous exclusion of such a feature. Robinson gives no details of the reasoning by which this structure was derived, citing only in a general way the now perceived difficulties with all previous structures, pointing out the structural relationship to the cinchona alkaloid cinchonine **48**, and promising that "a fuller discussion and description of relevant experiments will be submitted for publication elsewhere." As far as I have been able to determine, no such elaboration ever was published.

To say the least, the new structure must have caused surprise among the strychnine chemists. For example, after Woodward's lecture on the structure of strychnine at Columbia University in early 1947, a member of the audience asked his opinion of Robinson's new proposal. The response: "I regret to say that this must be a figment of his imagination."

In any case, the matter quickly dropped from view when a few months later, R.N. Chakravarti and Robinson submitted the structure **49** (Scheme 8.12) for strychnine. [43] This structure was one that had been considered by both Robinson [41] and Woodward [44] previously, but decisive evidence in its favor at first was not available. In very condensed form, Chakravarti and Robinson explained that conceptual difficulties hitherto preventing the acceptance of this structure now had been resolved. Specifically, they referred to the problem of the so-called neo-bases, of which neostrychnine, assigned structure **50** (Scheme 8.13), is the parent.

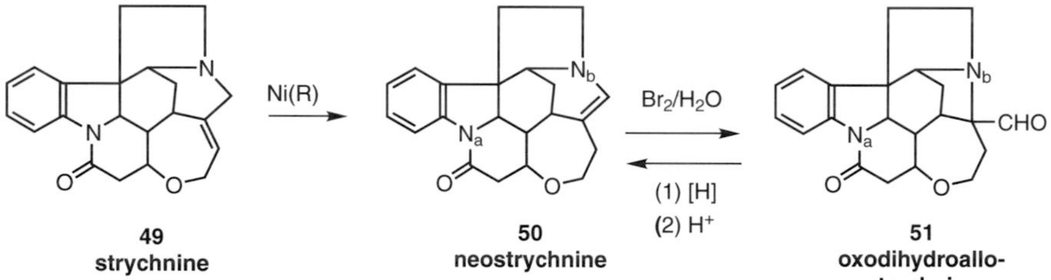

Ni(R)
Br_2/H_2O
(1) [H]
(2) H^+
49 **strychnine**
50 **neostrychnine**
51 **oxodihydroallo-strychnine**

Scheme 8.13

Neostrychnine is formed from strychnine in a number of ways, for example, by the action of Raney nickel on strychnine. Robinson had recognized that the most likely change in this reaction would be a shift of the double bond from the oxepine ring (VII) to the piperidine ring (VI). However, the properties of neostrychnine did not at first seem to be compatible with such a formulation. Most notably, no explanation was readily apparent for the formation of a base called oxodihydroallostrychnine when neostrychnine was treated with various oxidizing agents, most conveniently with bromine and water. The oxidation product contained a new carbonyl group but seemed still to contain the neostrychnine structure, since it could be reconverted to neostrychnine with acid. Eventually, Chakravarti and Robinson recognized that the new compound was formed by a contraction of ring VI of neostrychnine from six-membered to five-membered, accompanied by extrusion of the extra carbon in the form of an *aldehyde* function as in **51**. No doubt, the reader will be able to supply one or more straightforward mechanisms for this change. The aldehyde **51** can be reduced to a primary alcohol, which upon treatment with acid can be reconverted to neostrychnine **54**, thus bringing about a reverse rearrangement. Again, it is not hard to imagine how this might occur. It will be seen that the earlier strychnine structure **44**, with a five-membered N_b ring, would be required to form a strained four-membered ring in order to undergo the ring-contraction to an aldehyde. These observations convinced Robinson that strychnine was actually **49** instead of **44** (Scheme 8.14).

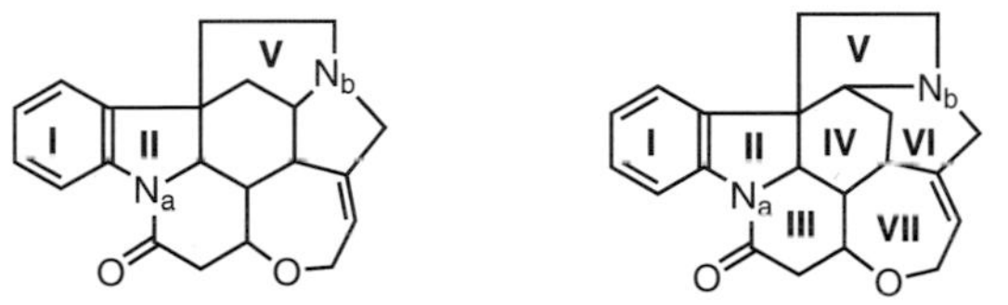

44 **49**

Scheme 8.14

In the following year, 1948, Woodward and Brehm [45] were to provide strong confirmatory evidence from other experiements for the same structure of neostrychnine. Before that, however, Woodward, Brehm, and Nelson, in a paper [46] which appeared in 1947 (submitted shortly after the Chakravarti-Robinson letter appeared), found an independent way to choose between the two strychnine structures **44** and **49** (Scheme 8.14). Leuchs and his co-workers [47–49] had discovered that the oxidation of pseudostrychnine (now formulated as **52**, Scheme 8.15) with acidic hydrogen peroxide gave a neutral substance *strychnone*, which they believed to contain a new carbonyl group.

The properties of strychnone, however, were puzzling. First, it did not give the so-called Otto reaction (a purple color produced by the addition of potassium dichromate to a solution of the base in 80% sulfuric acid). The Woodward group then observed [46] that the ultraviolet spectrum of strychnone, which showed $\lambda\lambda_{max}$ (log ε) 246 nm (4.15), 270 (3.02), 294 (3.72), was appropriate for an *N-acylindole*, not for an N-acyldihydroindole, such as strychnine itself. Consistently, N-acylindoles were known not to give the Otto reaction. They argued that strychnone was more readily formulated on the basis of structure **49** rather than **44** for strychnine. Thus, the changes could be understood readily as shown in Scheme 8.15. The key reactions were proposed to be the tautomerization of pseudostrychnine **52** to the aminoketone **53**, followed by what can be recognized as Baeyer-Villiger oxidation to the aminolactone **54** and aminolysis to the amide strychnone **55**. Note that these changes produce strychnone as a true fully aromatic indole.

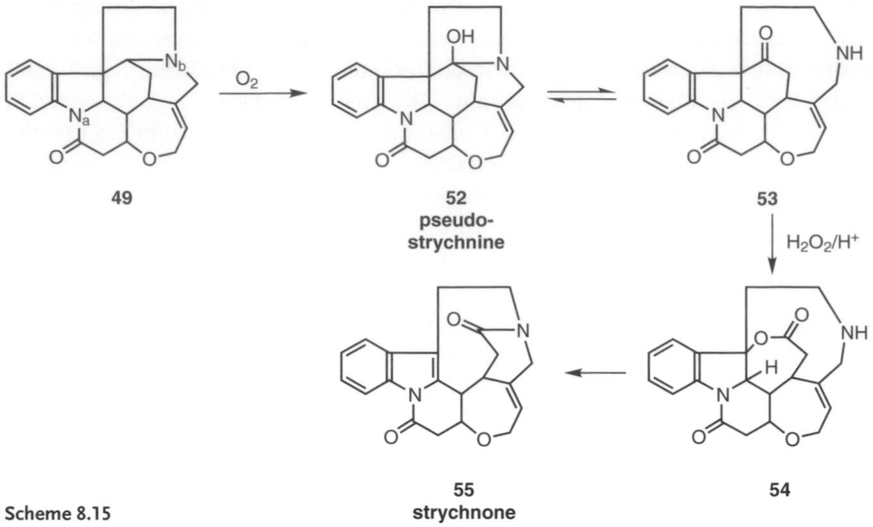

O_2 ; H_2O_2/H^+

49 **52 pseudo-strychnine** **53** **54** **55 strychnone**

Scheme 8.15

Woodward (Figure 8.2) and his co-workers interpreted this finding as "the long-sought crucial experiment with respect to resolving the remaining dubieties concerning the mode of linkage of N_b to the indole ring," an observation which showed that the β-position of pseudostrychnine's indole ring was connected to N_b by a single carbon atom, as in **49**, rather than by a two-carbon chain, as in **44** (Scheme 8.14).

A skeptic might point out that neither Robinson's nor Woodward's arguments were logically unassailable. For example, consider Robinson's claim to have shown that ring VI was six-membered, as in **49**, rather than five-membered as in the Menon-Robinson formula **44**. The claim was based upon the argument that if **44** were cor-

Figure 8.2 Robert Burns Woodward. Photograph contributed by Bernhard Witkop to the Othmer Archive of the Chemical Heritage Foundation.

rect, the ring-contracted aldehyde would have a four-membered ring. This is an *improbable* requirement, not a logically impossible one. Similarly, Woodward's claim to have shown the mode of attachment of N_b contained the unspoken assumption that a Baeyer-Villiger rearrangement would not be followed by a second rearrangement, for example, that shown in Scheme 8.16. There, the migration of the CH_2–CH_2 bridge to the alcohol methylene position is accompanied or followed by deprotonation to give ultimately **60**, a lactam with a fully aromatic indole ring.

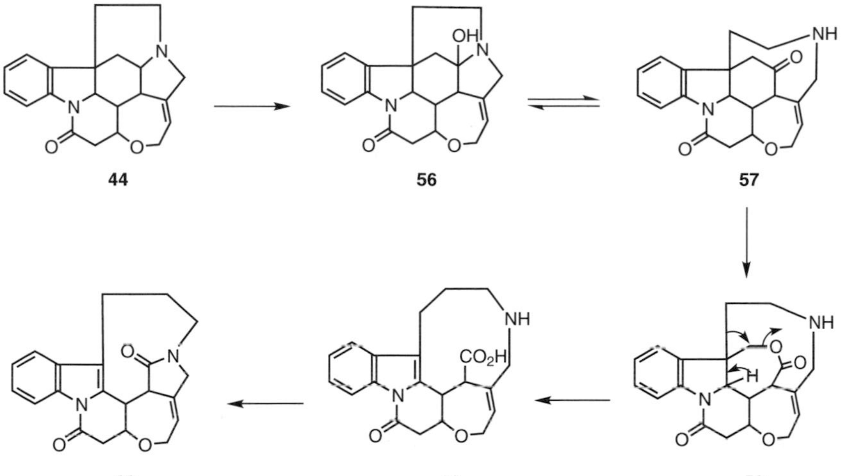

Scheme 8.16

Compound **60** would meet all the structural requirements of strychnone known at that time. Again, this rearrangement might be considered improbable, based upon an analysis of the two competing Baeyer-Villiger pathways shown in Scheme 8.17: In order to give a fully aromatic indole product via **58a**, Baeyer-Villiger migration of a methylene group must be favored over that of an allylic group leading to **58b**, whereas the reverse might have been expected on grounds of precedent. Nevertheless, a preference for **58a** is not logically impossible.

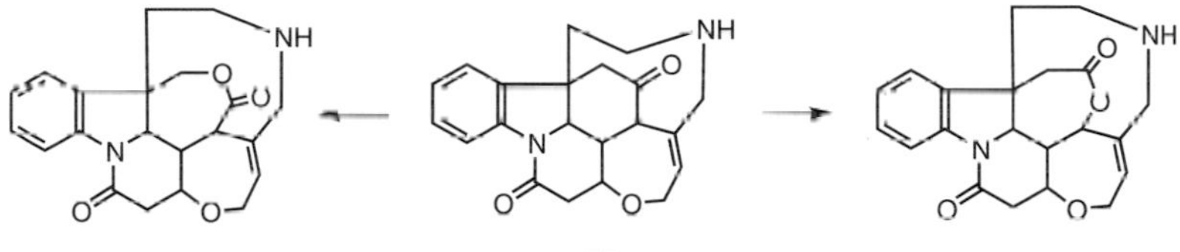

Scheme 8.17

Thus, the arguments of both Woodward and Robinson were based on the identification of improbable but not impossible alternatives. Importantly, however, they reinforced each other, because *both* would have to be wrong to permit formula **44** to be correct, and the probability of occurrence of *two* improbable events, which is just the product of their individual small probabilities, is very small indeed.

It is difficult to judge the exact level of conviction about this claim of proof that the organic chemical community held in 1947, but I sense that a certain reserve was still in the air. The skepticism was surely diminished when Woodward's confirmatory results became known, and the subsequent reports of single crystal X-ray structure determinations in 1949 and 1950, again confirmatory, followed by Woodward's brilliant total synthesis in 1954, [50–52] overwhelmed any remaining resistance.

8.5.4
Credit and Priority

In his review of the strychnine problem, [53] Robinson seems to make a point of claiming that his work on the final elucidation of the structure preceded all others. Thus, after summarizing his own findings of 1946 [41] and 1947 [43], he declares "the further evidence that later came from very different directions has all confirmed the correctness of **49** which must now be regarded as fully proved."

More recently, there has been a tendency for some authors to make the oversimplified formulation that Robinson proved the structure and Woodward did the synthesis. With regard to the synthesis, there is no argument. With regard to the structure, as we have seen, the work of many laboratories over many decades made significant contributions. In my view, the more accurate way to state the matter is the appropriately modest account given by Woodward [51] in a lecture at the International Congress of Pure and Applied Chemistry in 1955:

> "In this effort the work of one man, Sir Robert Robinson, shines brilliantly above that of all others. It was he who painted virtually the whole picture, basing his massive intellectual contribution in part upon the extensive experimental contribution of his own school, and making full use of the literally enormous wealth of painstakingly accumulated and beautiful experimental work of Hermann Leuchs and his collaborators. During the final stages of the campaign, some younger divisions were able to make small but substantial contributions to the taking of the last strongholds, and to help with the consolidation of the victory. Among these was that led by Professor Prelog here in Zürich, and I too had the good fortune to lead a small band of co-workers in Cambridge into the fray."

It may well be that Robinson felt pressured by Woodward's presence in the final hunt for the structure. After all, only a few years earlier Woodward had shown that Robinson's structure for penicillin was incorrect and deduced the correct beta-lactam structure. I was one of a group of students in Woodward's laboratory in 1949 who, in an informal setting, heard his account of the trip to England he had made a year or so before. Robinson had greeted him there with the question: "Dr. Woodward, why can't you work on your own problems?" Woodward, then very early in his career, seemed a bit shaken by Robinson's evident resentment. For a beginner, his encounters with this most famous of contemporary organic chemists were surely not the most diplomatic way to make a splash. Nevertheless, Robinson and Woodward developed a mutual personal respect and a high regard for each other's scientific ability. A case in point occurred when Pergamon Press, after launching the journal Tetrahedron in

1957, proudly proclaiming that it was "founded by Sir Robert Robinson," two years later launched the new companion journal Tetrahedron Letters. They chose Woodward, no doubt with Robinson's consent (or perhaps instigation), to serve with him as Co-Chairman of the Honorary Editorial Advisory Board. Considering Robinson's status and his concern to make sure that it was always appropriately acknowledged, his ready acceptance of Woodward into this position constituted a notable acknowledgment of collegial equality.

8.6 Speculations on the Biogenesis of Strychnine and Other Indole Alkaloids

The two-carbon connection of the indole β–position to N_b in the ultimately rejected structure **44** was an aspect of that structural hypothesis which worried both Robinson and Woodward and which made them reluctant to accept it. The concern was that it was difficult to reconcile with Robinson's biogenetic theory (see Section 8.1.4), a key feature of which was the Mannich condensation of an amine (either an amino acid or a derivative of it) with an aldehyde and a nucleophilic site. Thus, in the rejected structure **44**, one can discern a tryptamine unit, (Scheme 8.14, rings I, II, and the CH_2–CH_2 chain of ring V), but the two-carbon chain connecting N_b to the β-position of the indole ring clearly would have to come from some other biogenetic pathway. On the other hand, the ultimately correct structure **49**, with the single (presumably aldehydic) carbon connecting these two atoms, embodied the Mannich-based biogenetic pathway perfectly. The experimental vindication of structure **49** brought with it a soaring confidence in the overall validity of Robinson's biogenetic speculations.

In the previous decade, the Winterstein-Trier-Pictet-Robinson suite of biogenetic ideas had become prominent again as advances took place in the determination of the structures of several alkaloids. An important case was the isoquinoline alkaloid yohimbine **61** (Scheme 8.18) from *Corynanthe johimbe*, a crystalline base used as an adrenergic blocking agent in human medicine and also as an aphrodisiac in veterinary medicine.

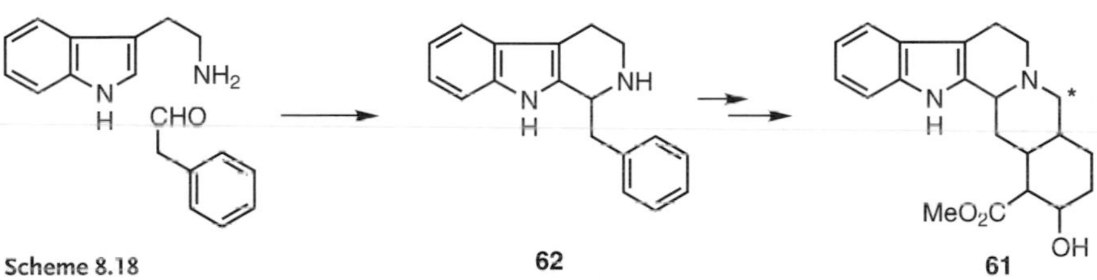

Scheme 8.18

Once the structure had been surmised, Barger and also Hahn pointed out [54;55] that the biosynthesis of yohimbine probably took place by reaction of one molecule each of tryptophan and phenylalanine (Scheme 8.18). The initial Pictet-Spengler condensation product **62** would require only the insertion of a "berberine bridge" (at the starred position in **61**) using a formaldehyde equivalent. Earlier, Robinson [19;23;56] had speculated that this type of condensation, which he imagined to occur by a Mannich reaction, was widespread in the biogenesis of alkaloids, prototypically in the case of berberine itself (**93**, see Scheme 8.26, Section 8.8.1). To complete the biosynthesis of yohimbine **61**, it would be necessary to make a few additional adjustments, the most significant of which included one that was to be recognized as a knotty problem: the origin of the carbomethoxy group (see Scheme 8.18). The Barger-Hahn scheme would require the ad hoc modification that this be derived from a one-carbon precursor in a separate pathway.

A significant further development of the Barger-Hahn hypothesis was Schöpf's application of Robinson's "physiological conditions" approach to the synthesis of isoquinoline alkaloids. [57]

8.6.1
The "Woodward Fission"

With the establishment of the structure of strychnine, speculation on its mode of biosynthesis could have been expected and indeed was not long in appearing. In 1948, [58] Woodward suggested a startling scheme which involved two new departures. The first differed in only a minor way from the previous proposals for yohimbine. Instead of condensation at the α–position of the indole nucleus of tryptamine **63**, Woodward envisioned condensation at the β–position (as in Scheme 8.19) with 3,4-dihydroxyphenylacetaldehyde **64** to give the intermediate **65**, in which the originally aldehydic carbon forming the single atom bridge to N_b is shown in bold. Insertion of a "berberine bridge" in the usual way and closure of another bond between the nucleophilic ortho carbon of the dihydroxybenzene unit and the electrophilic α–carbon of the indole ring would lead to **66**, which already embodies rings I, II, IV, V, and VI of strychnine **49**. Ultimate formation of Ring III could occur by aldolization of an acetic acid equivalent to a potential carbonyl group, but where does this group come from, and how can one understand the formation of the seven-membered Ring VII?

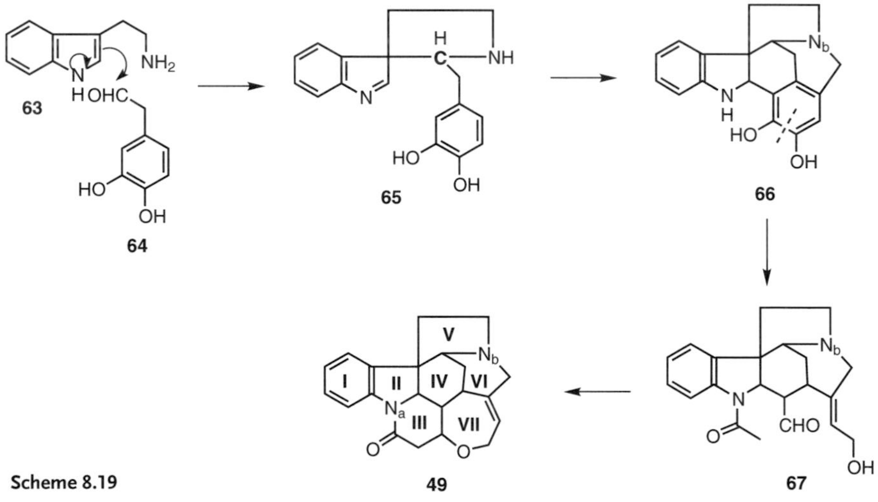

Scheme 8.19

Here Woodward made a second and much bolder departure from precedent. He proposed that the dihydroxybenzene ring at some stage would undergo cleavage to **67** (presumably, but not necessarily, oxidative) between the two hydroxy-bearing carbons. This process was later called by others a "Woodward fission." The resulting chains then would be free to form the seven-membered ether Ring III (Scheme 8.19). His triumphant conclusion is worth reading in full:

> "On the whole, the possibility of building up so complicated a structure as **49** by a series of simple reactions from plausible starting materials is so striking that it is difficult to believe that the scheme lacks significance. It is worthy of note that biogenetic considerations of this kind several years ago led us to consider that the environment of N_b in the strychnine molecule was that shown [in **49**] rather than that accepted at that time. We felt, however, that in view of the relatively undeveloped state of knowledge in the field of biogenesis, the proposal of these considerations as a possible biogenetic scheme should follow rather than precede the definite chemical proof of the primary consequences of its acceptance. Since this proof has recently been brought to a successful conclusion in our Laboratory, our confidence in the scheme is sufficient to bring us to suggest that other alkaloids will be found which owe their genesis to less, or more, complicated variants of the same process, and investigations in this direction are already under way."

In this passage, Woodward seems to have reached a state of Duhemian conviction that he had found the "natural classification." (see Chapter 2).

8.7 Speculations on the Biogenesis of Other Alkaloids

Woodward's idea of fission of a ring derived from a dihydroxyphenylalanine source had an immediate and enthusiastic reception. In a letter to *Nature* which was published as an appendix to Woodward's paper, [58] Robinson wrote:

> "When in New York in early 1947, I learned from Dr. Woodward that he had developed a scheme of biogenetic relations between strychnine and yohimbine. The details now available follow well-trodden paths in most respects; for example, the β–condensation would be included in any such hypothesis. The suggested fission of a benzene nucleus is, however, highly original. This daring speculation must be judged at this stage by 'the degree of coincidence' which called for it.
>
> In the thirty years that have elapsed since my first paper on these topics, considerations of this kind have been constantly in mind and conclusions have been reviewed and modified when necessary. I therefore feel justified in stating that in my opinion Woodward's theory is probably correct in essentials. May I also venture to deprecate in advance a flood of further speculation on similar lines, unless it is supported by a comparable 'degree of coincidence.' It is obvious that

by chopping up benzene rings and recombining the fragments, almost any kind of structure can be obtained."

Clearly, Robinson was agreeing with Woodward that the ring-fission interpretation of the biogenesis of the intricate strychnine structure was unlikly to be a mere coincidence.

8.7.1
Emetine

By 1948, work in several laboratories had reached the final stages of a structural assignment to emetine **72** (Scheme 8.20), the principal alkaloid of ipecac.

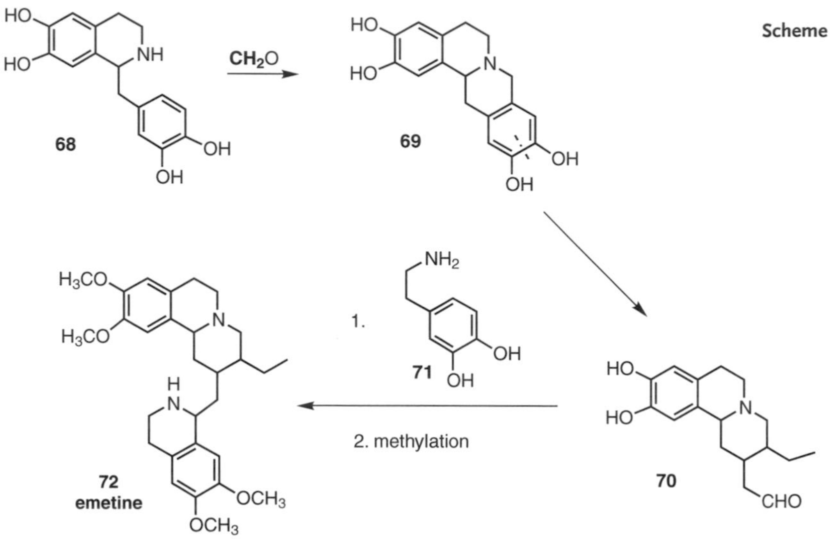

Scheme 8.20

Later in the same year [59] Robinson pointed out that, although he had not been following the literature on that problem, he was able, by invocation of the Woodward fission, to deduce a structure for a hypothetical alkaloid whose carbon-nitrogen-oxygen skeleton could be made up of three moles of dihydroxyphenylalanine and one mole of formaldehyde, a combination which, by loss of one NH_3 and three CO_2 and methylation of the hydroxyl groups would lead to structure **72**. Much to his surprise, he learned in a conversation with Michael Dewar that the recent experimental results pointed to exactly this structure as the most likely one for emetine. Thus, berberine-bridging of the familiar norlaudanosine structure **68** (the demethylated form of the of the Winterstein-Trier intermediate (see Scheme 8.6, Section 8.1) with formaldehyde could give the tetracyclic intermediate **69**. Woodward fission at the dashed line followed by adjustment of the oxidation state would give the aldehyde **70**, which is ready for another Pictet-Spengler reaction with dihydroxyphenylethylamine **71** to give, after methylation, emetine **72** itself. A key feature of this scheme is the very natural explanation of the presence of the ethyl side chain.

With enthusiasm tempered by caution, a review article by Birch, [60a] discussed the Woodward fission in the biosynthesis of strychnine and emetine as demonstrating

> "outstandingly elegant applications of biogenetic hypotheses...It is clear from the foregoing and other work that a series of consistent hypotheses exist which could now be tested in favourable cases by biological experiment using isotopic labels. The limitations of the approach are also clear: the hypotheses deal only with structural relations and structural origins in broad detail and give little idea of sequences of reactions."

8.7.2
Cinchona Alkaloids

Other chemists were quick to adopt the Woodward fission as an important element in the biogenetic classification of alkaloids. In 1950, Goutarel, Janot, Prelog, and Taylor [60b] proposed a biogenetic scheme (Scheme 8.21) for the cinchona alkaloids incorporating the fission as a key step.

Scheme 8.21

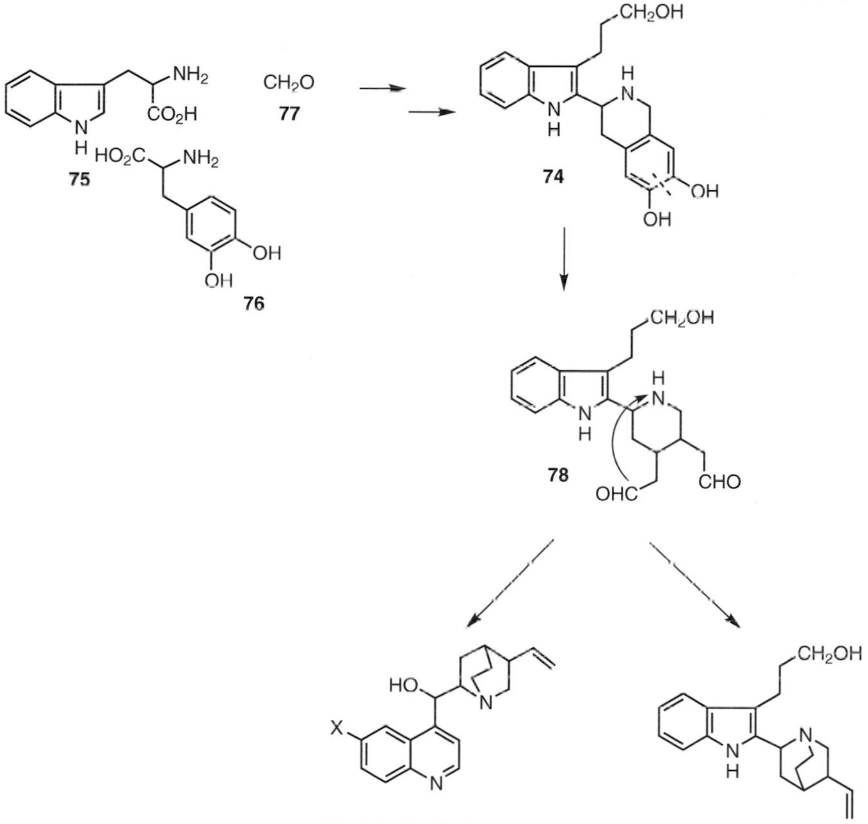

75 76 CH_2O 77 → 74 → 78 → 48a (cinchonine): X = H; 48b (quinine): X = OMe; 73 cinchonamine

These authors had established the structure of cinchonamine as **73**, an unusual base which is similar to the more common cinchona alkaloids cinchonine **48a** and quinine **48b** in containing a vinylquinuclidine residue but differs from them in the presence of an indole ring instead of the quinoline ring. In the authors' scheme, the intermediate **74** is formed by α–condensation of appropriately functionalized derivatives of tryptophan **75**, 3,4-dihydroxyphenylalanine **76**, and formaldehyde **77**. Woodward fission of **74** and cyclization at the piperidine nitrogen would give intermediate **78**, and adjustment of oxidation states would give cinchonamine **73**. The suggestion was made that cinchonamine might represent an early stage in the biosynthesis of the cinchona quinoline alkaloids cinchonine **48a** and quinine **48b**, which could be formed from the intermediate **78** by another new feature, expansion of the pyrrole ring. Although the authors did not specify a path for this transformation, it obviously could happen by cleavage of the pyrrole ring between the indole nitrogen and C_{α} followed by ring closure of the -CH_2CH_2OH chain at the indole nitrogen.

The paper concluded on an optimistic but again cautious note about such biogenetic speculations:

> "We might mention especially the heuristic value of such considerations, which permit several important alkaloid groups to be traced back to the same or similar building blocks. In the elucidation of structure, it can be easier in this way to choose the correct alternative from among several working hypotheses given by the experimental results. On the other hand it seems to us that our knowledge of the chemical processes *in vivo* is still too meager to complete a deeper discussion of the reaction mechanisms."

Despite caution from some, enthusiasm for the actual biogenetic significance of the fission hypothesis, not just its heuristic value, mounted throughout the 1950s. Thus, we find Turner and Woodward, in a review chapter [61] on the cinchona bases, stating unequivocally

> "it may be noted further that a similar cleavage is undoubtedly a factor in the formation of the ipecacuanha alkaloids, for example, emetine..."

In his Weizmann Lectures in 1953, [56] Robinson writes

> "Fantastic as Woodward's suggestion seemed at first sight, evidence of its correctness was soon forthcoming, and perhaps the most striking early development was the use that could be made of his hypothesis in an entirely different series to develop the structure of emetine."

8.7.3 Other Alkaloids

Woodward-fission origins were proposed during this period for a wide variety of other alkaloids from several plant genera, including (Scheme 8.22) the α–condensed indole alkaloids corynanthine **79**, [56] alstonine **80**, [56], and ajmaline **81.** [56;62;63]

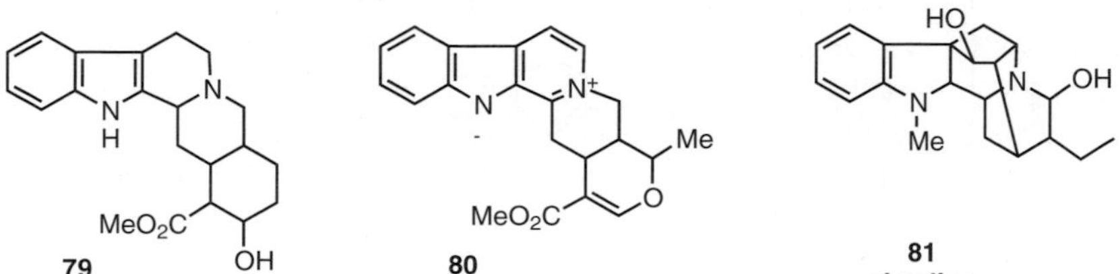

79 corynanthine

80 alstonine

81 ajmaline

Scheme 8.22

Robinson [56] was quite cautious about yohimbine **61** because of the problem we already have mentioned (Section 8.6), namely the apparent separate origin of the carbomethoxy group. He also made clear the need for efforts aimed directly at the biochemical mechanisms of phytosynthesis:

> "These and many other problems and speculations call urgently for the intervention of the plant physiologist armed with isotopes ... The organic chemist has all but completed the task of the determination of the molecular structure of plant products. The study of biogenesis in the future lies with the phytochemist, and in experiments in the laboratory of the living cell."

8.8
The Synthesis of Strychnine

Not only did the biogenetic hypothesis incorporating the Woodward fission (Scheme 8.19) offer a unifying overview of the natural origin of a broad range of alkaloidal structures, it also inspired Woodward's historic total synthesis of strychnine. [50–52;64] Thus, in order to force condensation at the indole β–position, in analogy to the proposed biosynthesis, Woodward chose to block the α–position with a veratryl group (see **82**, Scheme 8.23).

The intermediate **82** was cyclized successfully to the toluenesulfonylated amino ester **83**, which after reduction and acetylation gave the amide **84**. Moreover, in accordance with the biogenetic scheme, Woodward proceeded as follows: [51]

> "We now wished to modify the state of the relatively inert carbon atoms of the aromatic veratryl system in such wise as to permit their incorporation into yet further rings of the strychnine skeleton. To do this, it was clear in a general way that the aromatic ring must be broken."

In the original Woodward fission hypothesis, the procedure, oxidative or otherwise, by which the plant was imagined to achieve the cleavage of the six-membered ring (see **66** → **67**, Scheme 8.19) was not specified, but in the laboratory total synthesis, an actual oxidation was used, as ozone smoothly opened the veratryl ring of **84** (Scheme 8.23) to produce the muconic ester **85**. To describe in detail the steps used in the conversion of the latter to strychnine **49** would divert us from our purpose, but it should be clear already that in these few bold but carefully planned early steps,

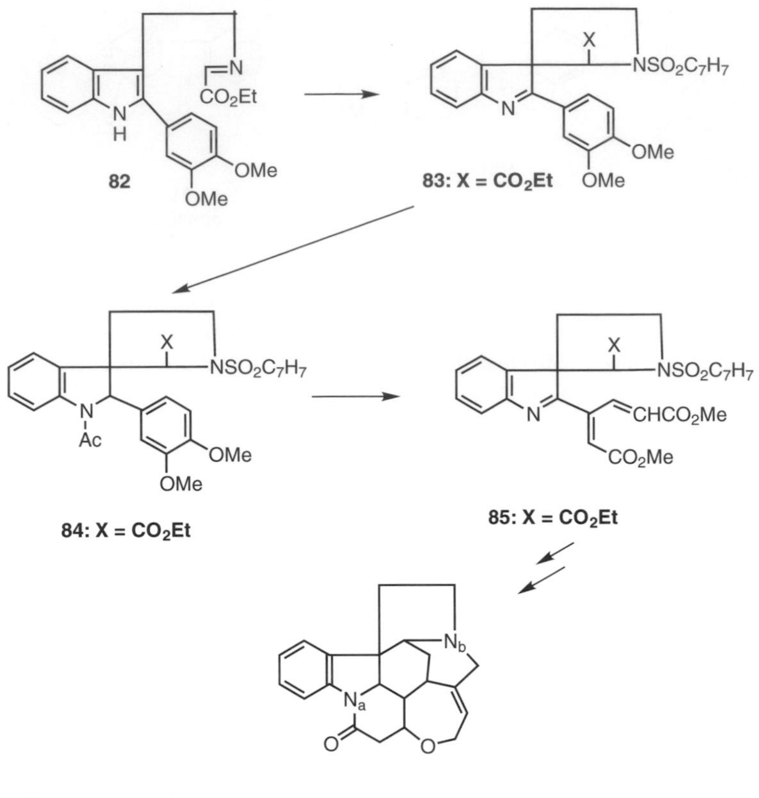

Scheme 8.23

Woodward and his co-workers had constructed almost all of the carbon-nitrogen skeleton of strychnine. The analogy to the imagined biogenetic pathway is striking. Moreover, Woodward notes that the selective ozonolysis of a benzene ring between two oxygenated positions

> "does not appear to have commanded attention; it seems not unlikely that it might provide a route to a rich variety of reactive and otherwise difficultly accessible intermediates from readily available aromatic compounds."

I cannot leave this account of one of the inspiring stories in the history of organic chemistry without pointing out that Woodward shared the sense of awe mixed with joy that his audiences felt in the presence of this synthesis, emotions that he expressed only in playful form. Thus, in describing the optical resolution of an intermediate in the synthesis, he remarks, "We will admit to having taken a certain humorous pride in our use of quinidine as resolving agent. For, some years previously, this alkaloid too had been synthesized in our Laboratory [with W. von E. Doering in 1944], and at the point in that work at which a resolution was necessary, dibenzoyltartaric was utilized. If it be recalled that tartaric acid had been resolved manually in the form of its sodium ammonium salt [by Pasteur in 1848], we can claim that our total synthesis [of strychnine] has been achieved without the intervention of any natural agent, other than one which, happily, is not likely ever to be completely eliminated from organic chemical investigations. In a similar vein, we may note here the pleasure taken in the outcome of our work on the part of our good friend Dr. Sarett, whose use of strychnine for the resolution of an intermediate in the course of his brilliant work on the synthesis of cortisone, left his investigations, in the jocular sense used here, unfinished until now."

At risk of the accusation of quibbling, we might point out that the "natural agent" to whom Woodward referred, namely the chemist, is of course chiral and non-racemic, just as the resolving agents are. The difference is merely a matter of definition.

An interesting question remains: As we have seen, Woodward presented his fission hypothesis of the biogenesis of strychnine as an idea *de novo* (see Section 8.6.1). Taken at face value, this implies that the biogenetic scheme inspired the actual laboratory synthetic route. What we don't really know is whether this impresssion is correct, or whether the reverse is true, namely that he might have been thinking of ways to synthesize strychnine *in vitro*, conceived the fission step, and *then* realized that this might be a plausible biogenetic pathway. Nothing in his own account of the events actually states that this was the case, but as we have seen in connection with the Robinson tropinone synthesis (Section 8.1.4), the actual sequence of such purely cognitive events can be difficult for an observer (and perhaps even for the investigator) to establish.

8.9 The Experimentally Determined Biogenesis of Alkaloids

8.9.1 Benzylisoquinolines and Pyrrolidines

Although the early speculations of Pictet, Winterstein, Trier, and Robinson were put forward in the first two decades of the 20th century, and were repeated and refined by Robinson during the ensuing years, there actually was little experimental evidence for even the most fundamental feature of the reigning biogenetic hypothesis, namely that the building blocks are amino acids, which combine with other simple compounds such as acetate, formaldehyde (possibly derived from the amino acid glycine) and in some cases, biological methylation agents. We recall (see Section 8.1.1) that Pictet imagined the amino acids to be derived from the breakdown of proteins, but more generally, it was not really known that amino acids at any stage of metabolism were the precursors of alkaloids.

In the last thirty years or so, much has been learned from isotopic tracer experiments about the actual biosynthetic pathways. In some cases, the early theories were shown to be correct, at least in the broad sense. For example, in the benzylisoquinoline group, two equivalents of ^{14}C-tyrosine **86** (Scheme 8.24) are incorporated into both laudanosine **16** and papaverine **30** in the opium poppy *Papaver somniferum*, [65] in accord with the Winterstein-Trier hypothesis (see Scheme 8.6). Similarly, Robinson's suggestion [19] that ornithine (2,5-diaminopentanoic acid **87**) is a precursor of hygrine **88** and the tropane alkaloids is consistent with the finding [66] that labeled ornithine **87** (Scheme 8.25) is incorporated in the plant into hygrine **88** and cuscohygrine **89**.

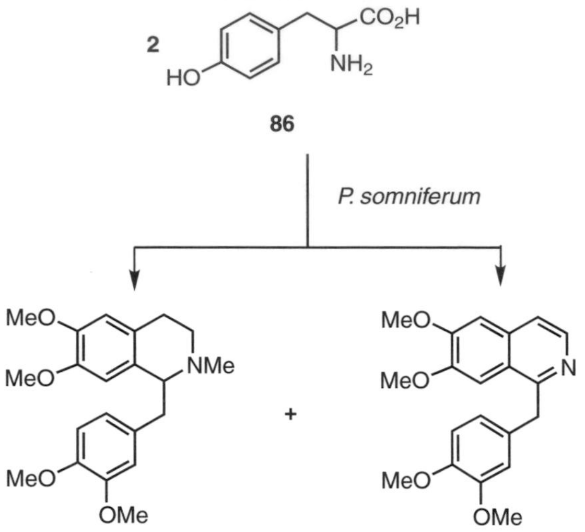

2 HO CO_2H NH_2

86

P. somniferum

MeO MeO NMe MeO OMe + MeO MeO N MeO OMe

16 **laudanosine** **30** **papaverine**

Scheme 8.24

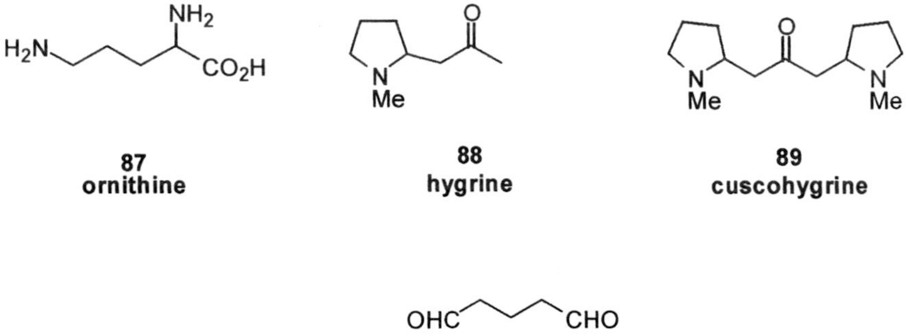

H_2N NH_2 CO_2H O N Me O N Me N Me

87 **ornithine** **88** **hygrine** **89** **cuscohygrine**

OHC CHO

succindialdehyde **20**

Scheme 8.25

However, it was shown that 2-^{14}C-labeled ornithine is incorporated in such a way that C-2 and C-5 do not become equivalent, that is, without passing through a symmetrical intermediate. [66] Thus, Robinson's scheme for the formation of these bases from ornithine **87** by decarboxylation and oxidation of the latter to succindialdehyde

20, followed by Mannich condensation, in analogy to the laboratory synthesis of tropinone **19** (see Scheme 8.7), seems to be oversimplified.

A further modification of the Robinson scheme became necessary when it was shown [67;68] that the true biosynthetic pathway to the so-called "berberine bridge" carbon of berberine itself **93** (Scheme 8.26) does not involve a Mannich reaction of a "formaldehyde equivalent," as Robinson had speculated. [19;23;56] It is true that this carbon is derived from the one-carbon pool, since ^{14}C-formic acid is incorporated, but the mechanism does not involve a Mannich reaction. Instead, the berberine bridge site (*) apparently is derived from the N-methyl group of the tricyclic precursor reticuline **90** (Scheme 8.26).

Scheme 8.26

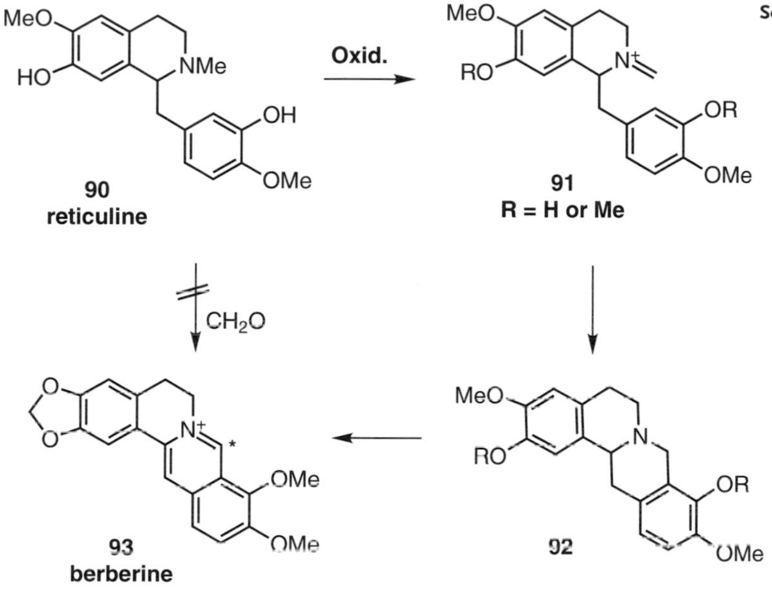

Notice that nature has played a subtle trick here. The cyclization step from the iminium ion **91** to the intermediate **92** is in fact a Mannich reaction, but a most peculiar one. The iminium ion **91** is not derived by a conventional Mannich dehydration of a carbinolamine, as Robinson had envisioned, [19;23;56] but rather by *oxidation* of an N-methyl group! Surely he can be forgiven for the incorrect speculation of nearly fifty years before. Indeed, it is astonishing that he came so close to the right answer.

8.10
Tests of the Biosynthetic Role of the Woodward Fission in the In Vivo Formation of Strychnine and Other Alkaloids

As we have seen, the concept of the Woodward fission of a phenylalanine- or tyrosine-derived dioxyphenyl ring as applied to the indole alkaloids, *e.g.*, strychnine

(Scheme 8.19, **49**), corynanthine (Scheme 8.22, **79**), alstonine (Scheme 8.22, **80**), and ajmaline (Scheme 8.22, **81**), to emetine (Scheme 8.20, **72**), and to the cinchona alkaloids, *e.g.*, cinchonine and cinchonamine (Scheme 8.21, **48a** and **73**), had contributed to outstanding progress in the structural relations of natural products and in synthesis.

Again, however, there was little or no hard evidence that plants actually employed that pathway in building up the alkaloids. Some workers in the field had expressed such reservations, but the protagonists of the theory, with infectious enthusiasm, insisted that this was not only a useful organizational device but biochemical reality. Events were to prove otherwise.

8.10.1 Alternatives to the Woodward Fission

By the mid-1950s, some specific counter-proposals began to appear. These were stimulated at least in part by the exciting developments in biochemical mechanistic research which revealed the actual pathways of biosynthesis [69] of such important materials as amino acids, steroids, and terpenes. For reasons to be described, in the period 1959–1961, speculations arose that the non-indolic portions of many alkaloids might not be derived from the Woodward fission of an aromatic ring precursor, but instead might have a completely different origin. By the 1960s and 70s, as the result of the work of several laboratories, compelling experimental facts incompatible with the Woodward fission hypothesis came to light and showed that the actual pathways were indeed entirely different. These events have been reviewed repeatedly, and the reader is referred to those sources [70–77] for details and references to the primary literature. The purposes of the current discussion will be served by a brief outline of the early history.

Ernest Wenkert was among the more outspoken skeptics of the reigning biogenetic theory of the alkaloids. One of the problems that led him to question the hypothesis of an amino acid origin of the entire carbon skeleton of alkaloids was the frequent occurrence in nature of important portions of some alkaloid structures that appeared to have been derived structurally from acetate or terpenoid precursors, rather than from amino acids such as tyrosine. To understand how drastic a revision this proposal was, one has to remember a few facts about the biosynthesis of some simple building blocks (see Scheme 8.27).

There are three major streams that feed the biogenetic pool and are potentially relevant to our concern here. These are: Pathway I, originating in acetate and then mevalonate and ending up in steroids, terpenes, and iridoids; pathway II, also originating in acetate and proceeding on through malonate to fatty acids; and pathway III, originating in fructose, passing through shikimic acid and chorismic acid and then branching. Branch IIIA goes to tryptophan and ultimately into the indolic portion of many alkaloids. Branch IIIB goes to tyrosine and other aromatic amino acids. It should be obvious from our previous discussion that the biogenetic proposals of Robinson and of Woodward made use of the amino acid products of Pathways IIIA and IIIB, respectively, as the sources of both the indolic and non-indolic moieties of

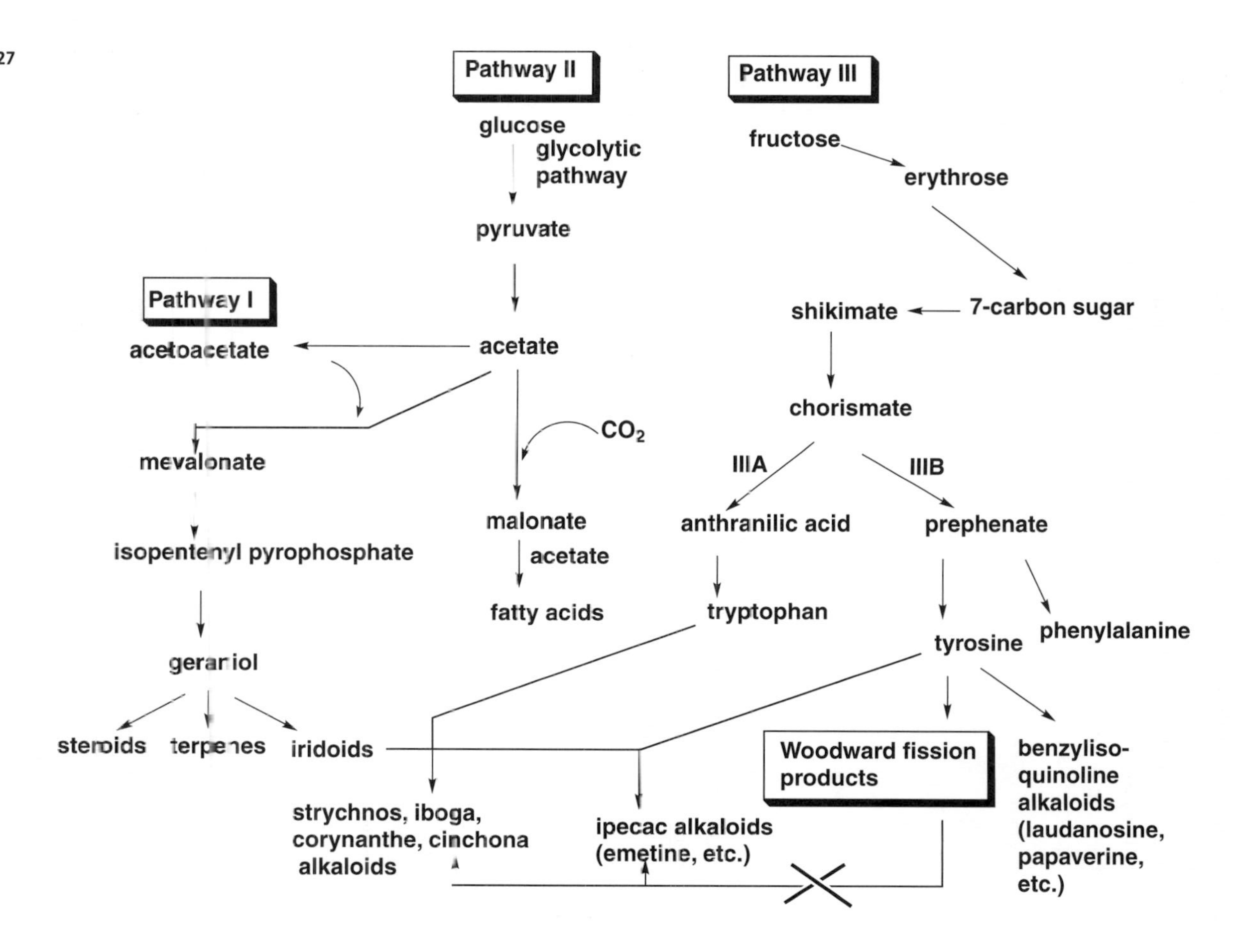
Pathway II
Pathway III
glucose
glycolytic
pathway
fructose
erythrose
pyruvate
Pathway I
acetoacetate
acetate
shikimate
7-carbon sugar
chorismate
CO_2
mevalonate
IIIA
IIIB
malonate
anthranilic acid
prephenate
acetate
isopentenyl pyrophosphate
fatty acids
tryptophan
tyrosine
phenylalanine
geraniol
steroids
terpenes
iridoids
Woodward fission
products
benzyliso-
quinoline
alkaloids
(laudanosine,
papaverine,
etc.)
strychnos, iboga,
corynanthe, cinchona
alkaloids
ipecac alkaloids
(emetine, etc.)

Scheme 8.27

a large group of alkaloids. What was now emerging was entirely novel idea that part of the carbon skeleton of alkaloids comes instead from Pathway I, the acetate-mevalonate-terpenoid-iridoid route (Scheme 8.27).

At one time, Wenkert and Bringi, [78] by a long and not entirely compelling argument, suggested that shikimic acid **94** (Scheme 8.28), a biogenetic precursor of phenylalanine **95**, might be the progenitor of the non-indolic portion of indole alkaloids. They imagined that the plant's synthetic machinery incorporated shikimate by intercepting it on the pathway to, but short of actual arrival at, the aromatic amino acids phenylalanine and tyrosine (see Scheme 8.27).

Scheme 8.28

94 shikimic acid

95 phenylalanine

At this stage, Wenkert and Bringi [78] had not yet made a clean break from the Woodward fission hypothesis, since the new proposal, like his, still envisioned a biogenesis starting on Pathway III. Another point of similarity was the requirement for a ring-opening at some point to construct rings D and E of yohimbine. The authors suggested that this might occur by a retro-aldol reaction on an alicyclic intermediate.

8.10.2 The Wenkert-Thomas Hypothesis

These ideas, however, soon were replaced, when Wenkert [79] and Thomas [80], independently and almost simultaneously, put forward a new hypothesis which abandoned the Woodward fission once and for all. The new theory differed drastically from any previously advanced, because it proposed that the biosynthesis was initiated not in Pathway III but instead in Pathway I. These proposals were based on an inspection of the structures of several non-nitrogenous natural products called iridoids (Scheme 8.29) discovered in other laboratories, including verbenalin **96**, genepin **97**, aucubin **98**, and asperuloside **99**, all of which seemed to be related to a common cyclopentanoid unit **100** whose exact functionality was unspecified.

It then took an intuitive leap of the imagination to propose that cleavage of **100** as shown in Scheme 8.30 could give an intermediate with skeleton **101**, which could recyclize in another direction to an intermediate with skeleton **102** or could rearrange to another one with skeleton **103**.

Combination of such a C_{10} terpenoid species with tryptamine or tyrosine in various ways could account for the major structural elements of most of the indole alkaloids and other bases that had been imagined previously to arise via Woodward fission. Scheme 30 outlines examples of speculative biogenetic paths to the formation of the D and E rings of yohimbine **61** as well as to the non-indolic portion of aspidospermine **104**, emetine **72**, and strychnine **49**.

Scheme 8.29

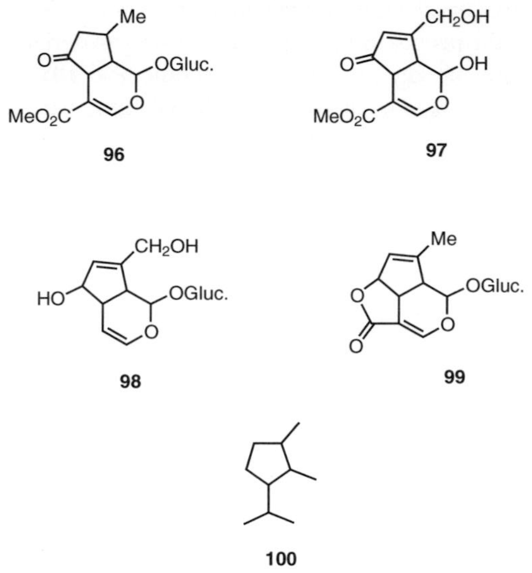

96 **97**

98 **99**

100

Scheme 8.30

100

101

103

72
emetine

104
aspidospermine

49
strychnine

102

61
yohimbine

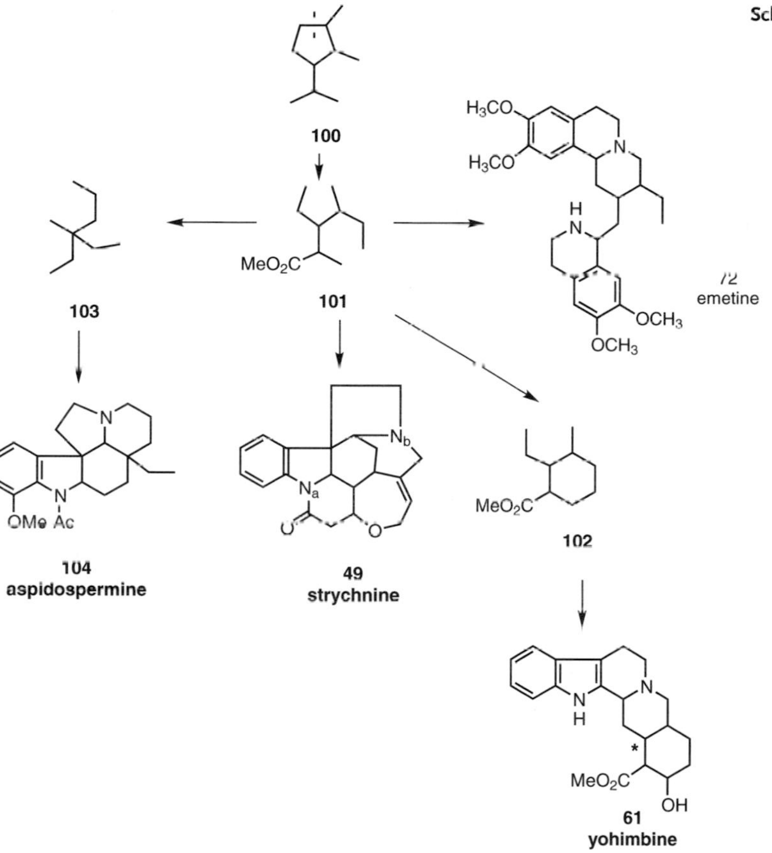

By 1960, the main outlines of the mechanisms of terpene biosynthesis had been established [69] (see Scheme 8. 27). If the proposed iridoid species **100** were derived by such pathways, it should arise from the key precursors mevalonic acid **105** or mevalonolactone **106** (Scheme 8.31). This was soon confirmed by labeling experiments on a number of alkaloids.

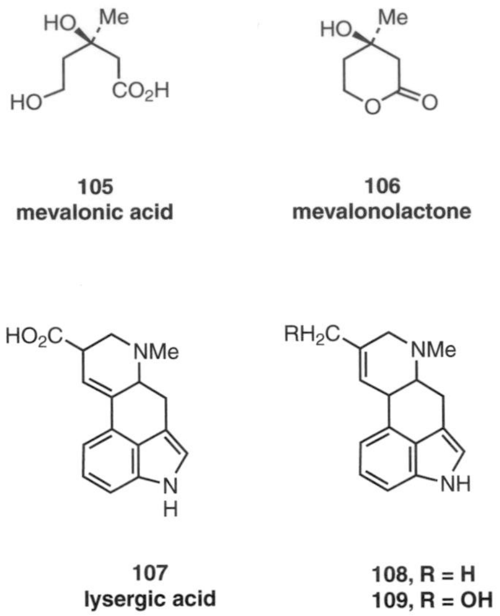

105
mevalonic acid

106
mevalonolactone

107
lysergic acid

108, R = H
109, R = OH

Scheme 8.31

Actually, however, the alkaloids around which the Woodward and Wenkert-Thomas schemes had been developed were not involved in the first demonstration of a terpenoid origin of part of an alkaloid skeleton. Rather, the structurally quite different alkaloids of the ergot family were the first. These bases **107–109**, the best known of which is lysergic acid **107**, (Scheme 8.31) are derived from the fungus *Claviceps purpurea*, which grows as a parasite on rye and other grasses. In their structures, one can discern a tryptamine unit plus a *five-carbon* unit, instead of the C_{10} unit characteristic of strychnine, emetine, and the other related alkaloids derived from higher plants. Even before the Wenkert-Thomas proposals, other workers [81;82] had speculated that the ergot alkaloids might be derived from tryptamine and one molecule of the five-carbon terpene precursor mevalonate. In 1960, Birch and co-workers had shown experimentally that the indolic portion indeed was derived from tryptamine, a point which was not really in serious question, but also, by exposing growing *Claviceps* cultures to(2-^{14}C)-mevalonolactone **106** or to (appropriately labeled) acetate ion, its biochemical precursor, that the positions of the radioactivity incorporated were just those required by a mevalonoid origin of the five-carbon unit of the ergot alkaloids agroclavine **108** and elymoclavine **109** [82].

Soon after the Wenkert and Thomas proposals appeared in 1961–2, isotopic labeling experiments designed to test the relative merits of the Woodward fission hypothesis and the Wenkert-Thomas iridoid-terpenoid hypothesis were undertaken. Again, the results were in decisively better agreement with the iridoid-terpenoid scheme for a number of alkaloids, including in the early stage [70;76] several members of the indole group (ajmalicine **110**, catharanthine **111**, serpentine **112**, and perivine **113**, Scheme 8.32), and subsequently other members of the Woodward fission "family," including emetine **72** [83;84] and strychnine **49** [85]

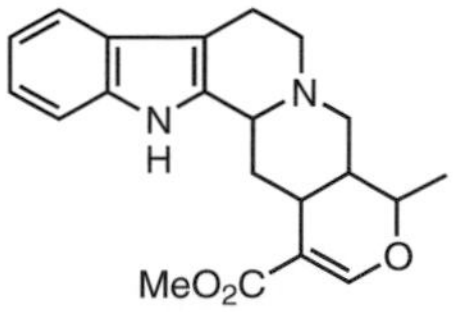

110 ajmalicine

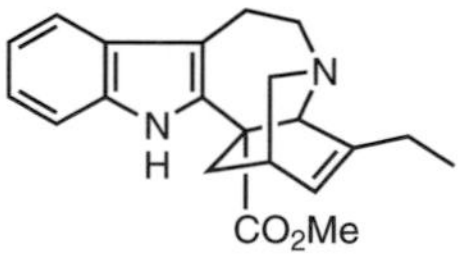

111 catharanthine

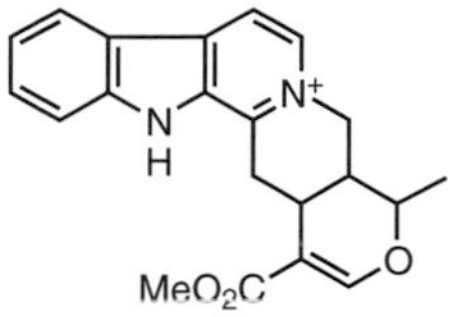

112 serpentine

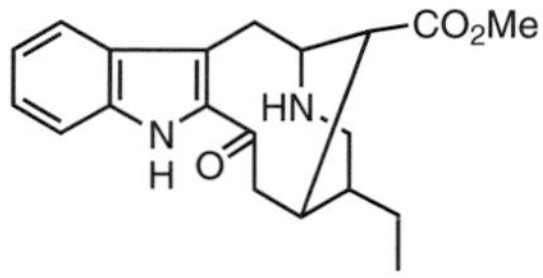

113 perivine

Scheme 8.32

Several laboratories reported the incorporation of labeled mevalonate into alkaloids of the latter type and also demonstrated the intermediate stage of C_{10} terpenoid biosythesis in the corresponding plant species by incorporation of geraniol (**114**, Scheme 8.33), as is described in reviews. [70;76]

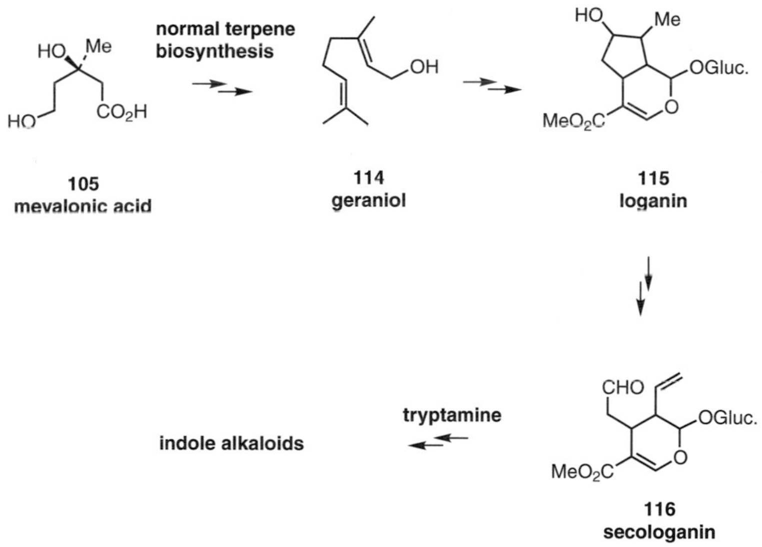

normal terpene biosynthesis

105 mevalonic acid

114 geraniol

115 loganin

tryptamine

indole alkaloids

116 secologanin

Scheme 8.33

Perhaps the most telling experiments establishing the later stages of the biosynthesis were those which clarified the structure of the actual C_{10} terpenoid building blocks common to all these alkaloids. We recall that Wenkert and Thomas had speculated that the naturally occurring iridoids, such as verbenalin **96** and genepin **97** (Scheme 8.29) or other related species might serve this purpose. In fact, however, direct experimental tests with isotopically labeled verbenalin **96** or several other related molecules showed that none of them is incorporated significantly into the alkaloids of *Vinca rosea*, such as catharanthine **111** and perivine **113** (Scheme 8.32). [70]

8.10.3
Loganin and Secologanin, Key Intermediates from Mevalonate, not from Amino Acids

However, in 1961 two research groups [86] reported new structural results on another C_{10} terpenoid, loganin. Two facts made this compound interesting in connection with the biogenetic problem: first, its structure **115** (Scheme 8.33) [86] closely resembled those of the iridoids **96–99** (Scheme 8.29), and second, it had been isolated [87] (almost 80 years before!) from *Strychnos* species. This biological origin was a suggestive hint that loganin might be the missing link in the biosynthesis of the *Strychnos* alkaloids and, by extension, perhaps of such other bases as emetine, ajmalicine, etc.

Massive experimental efforts over a period of years in many laboratories, as reviewed, [71;73;74;76] followed up on this hint. We can summarize the results with the statement that loganin **115** is indeed a key intermediate on the pathway to many of those alkaloids. It is converted by a ring-opening reaction *in vivo* to a later intermediate secologanin **116** (Scheme 8.33). Both loganin and secologanin are now known to be actively incorporated into many alkaloids in growing plants. Thus, the major points along the biosynthetic pathway to the indole alkaloids now have been identified as shown in Scheme 8.33. It is clear that the idea of an iridoid origin of key portions of many alkaloids the Wenkert-Thomas hypothesis, given in skeleton form in Scheme 8.30, has been verified. Secologanin **116** embodies the main stuctural elements of the acyclic structure **101** of the Wenkert-Thomas scheme and differs from it only in the detailed specification of the functionality, as is easily seen in Scheme 34. Many of the the detailed biochemical steps now have been further elucidated to give almost complete corporality to the mechanisms.

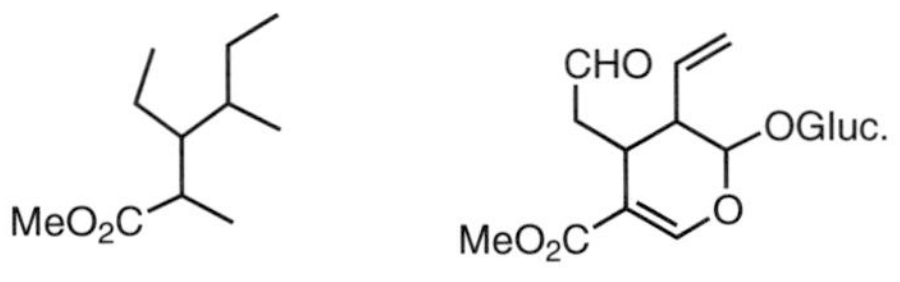

MeO$_2$C

101
hypothetical iridoid intermediate

CHO OGluc. O MeO$_2$C

116
secologanin

Scheme 8.34

8.11
The Startling Accuracy of a False Theory

Woodward's proposal was that the non-indolic portions of the indole alkaloids arose *in vivo* from phenylalanine or its derivatives such as tyrosine or 3,4-dihydroxyphenylalanine. He imagined that, after incorporation into the biosynthetic sequence, the aromatic ring was cleaved between two carbon atoms to make available two side-chains for further elaboration of the alkaloid structure. It was one of his characteristically bold and brilliant concepts. Woodward fission was the dominant idea guiding structural elucidation and synthesis of alkaloids for almost twenty years. Its heuristic value can be appreciated if we realize that its ultimate overthrow did not necessitate any significant revisions of alkaloid structures deduced in its name. In other words, it was a powerfully accurate *predictive* tool.

We might ponder how this could have been so. Was it a matter of sheer coincidence? A moment's thought reveals the reason. Woodward's hypothesis really consisted of two parts, one structural and the other biochemical. The structural part was a deep analysis of atomic connectivity in the alkaloids, starting, as we have seen, with the understanding of the diabolically intricate strychnine molecule. This analysis revealed the *structural patterns* that were repeated, even if in masked form, from one alkaloid to another. Once this insight had been gained, *any hypothesis* capable of reproducing these regularities would serve a structurally predictive purpose.

The second part of the Woodward fission hypothesis was the actual proposal that these structural insights had their roots in a particular biochemical mechanism, namely that the problematical non-indolic portion of the indole and ipecac alkaloids is derived by cleavage of the aromatic ring of tyrosine or a derivative thereof.

However, what is now abundantly clear is that, contrary to Woodward's hypothesis and to its extensions by Robinson and (cautiously) by Prelog, tyrosine plays no significant role in the actual biosynthesis of those alkaloids. Instead, it is the iridoids loganin **115** and secologanin **116** (Scheme 8. 33), derived from mevalonate (Pathway I of Scheme 8.27), that provide the carbon skeleton of those molecules.

8.11.1
Popper on Replacement of a Theory

At this point, it can be instructive to recognize that the history of alkaloid biogenesis we have just surveyed provides an example of another of Popper's dubious characterizations of how science works. This comes from an examination of his proposals for the relationship between a refuted theory and the one that replaces it. In his words, [2]

> "One may discern something like a general direction in the evolution of physics – a direction from theories of a lower level of universality to theories of a higher level....a theory which has been well corroborated can only be superseded by one of a higher level of universality; that is, by a theory which is better testable and which, in addition, *contains* the old, well-corroborated theory – or at least a good approximation to it."

To say that theory A "contains" theory B suggests that B can be derived from A or that B predicts the same result as A under certain specified conditions. These are not features that can reasonably be said to pertain to the two theories here.

Consider the events just recounted in which the well-corroborated Woodward fission theory was replaced by the Wenkert-Thomas iridoid theory. In this case, as in many others, it is not a question of the old theory being less universal. Sometimes the old theory is simply wrong. Does the Wenkert-Thomas theory "contain" the Woodward theory? It is true that, in a loose sense, one could say that relics of Woodward's theory survive in the iridoid theory, which shares with it the feature that a carbocyclic ring must be cleaved to produce the actual biosynthetic intermediates. But the hypothetically cleaved rings in the two theories are of different sizes (6-membered in the Woodward theory and 5-membered in the Wenkert-Thomas theory), and this is for the good reason that they arise from completely disparate biochemical pathways.

8.12
Dreamers and Practitioners

One of the epistemological lessons to be learned from the story of the Woodward fission is that some theories, despite deep flaws which cannot be corrected within their own framework, nevertheless organize a great body of empirical fact and accurately predict the outcome of new experiments. During their lifetimes, they are not only useful but also inspirational. In this respect, there is a parallel between Woodward's proposal and other transient hypotheses, like the Bohr theory of the atomic spectral lines, which we discussed in Chapter 3, and the Kekulé collision theory of chemical bonding, which we discussed in Chapter 4. These experiences teach us that we must not be afraid to speculate. When a field is new and undeveloped, someone must have the nerve to propose new and unprecedented ideas. Someone must have the self-confidence to risk eventually being shown wrong. Science must value the dreamer as well as the hard-headed practical investigator. Few of us dare to dream, so we must be grateful for the Woodwards, the Bohrs, and the Kekulés.

8.13
References and Notes

1 R. P. Feynman, *The Nature of Physical Law*, MIT Press, Cambridge, MA, **1965**. p. 156
2 K. R. Popper, in *Conjectures and Refutations: The Growth of Scientific Knowledge.*, Routledge and Kegan Paul, London, **1963**. p. 276. Reprinted from the revised fifth edition of 1989 by Routledge, 1996.
3 J. A. Berson, *Chemical Creativity: Ideas from the work of Woodward, Hückel, Meerwein, and Others.*, Wiley-VCH, Weinheim, Germany, **1999**. Chapter 4.
4 A. Pictet, *Arch. Pharm.* **1906**, *244*, 389 and references cited therein.
5 G. Barger, *J. Chem. Soc.* **1938**, 1113. A biographical memoir.
6 E. Cherbuliez, *Helv. Chim. Acta* **1937**, *20*, 828. A biographical memoir.
7 A. Pictet, T. Q. Chou, *Ber.* **1916**, *49*, 376.
8 D. R. Dalton, in *The Alkaloids: The Fundamental Chemistry – A Biogenetic Approach*, Marcel Dekker, New York, **1979**. p. 147ff. Also references cited therein.

9 G. A. Cordell, in *Introduction to Alkaloids: A Biogenetic Approach*, Wiley-Interscience, New York, **1981**. p. 94ff.
10 A. Pictet, T. Spengler, *Ber.* **1911**, *44*, 2030.
11 B. Reichert, in *Die Mannich Reaktion*, Springer, Berlin, **1959**. p. 1–3.
12 F. F. Blicke, in, in *Organic Reactions*, Wiley, New York, **1942**. I p. Chapter 10.
13 H. Schaefer, B. Tollens, *Ber.* **1905**, *39*, 2181.
14 C. Mannich, *Arch. der Pharm.* **1917**, *255*, 261. This introduced a series of articles in following years exploring the scope of the reaction. However, it was not Mannich's first paper on the condensation of amines, aldehydes, and active hydrogen compounds: See Mannich, C.; Krösche, W. *Arch. Pharm.* **1912**, *250*, 647.
15 C. Mannich, W. Krösche, *Arch. Pharm.* **1912**, *250*, 647.
16 E. Winterstein, G. Trier, *Die Alkaloide*, Gebrüder Bornträger, Berlin, **1910**.
17 A. J. Birch, *Notes Rec. R. Soc. London.* **1993**, *47*, 277.
18 R. Robinson, *J. Chem. Soc.* **1917**, *111*, 762.
19 R. Robinson, *J. Chem. Soc.* **1917**, *111*, 876.
20 T. A. Henry, in *The Plant Alkaloids*, Blakiston, Philadelphia, **1949**. p. 64ff.
21 L. Small, in *Organic Chemistry, An Advanced Treatise*, 2nd ed. (Ed.: H. Gilman) John Wiley, New York, **1943**. II p. 166.
22 R. Willstätter, *Ann.* **1903**, *326*, 23.
23 R. Robinson, *J. Chem. Soc.* **1936**, 1079.
24 Robinson was not insensitive to appropriate recognition of priority, but his zeal for it seemed to increase when he felt his own priority had been slighted. Prominent among the examples is his bitter dispute with Ingold over the origin of the electronic theory of organic chemistry (see the following reference).
25 R. Robinson, in *Memoirs of a Minor Prophet: 70 Years of Organic Chemistry*, Elsevier, Amsterdam, **1976**. v. 1 p. 73–76.
26 A. Pictet, A. Gams, *Ber.* **1909**, *42*, 2943.
27 W. H. Perkin, Jr., R. Robinson, *Proc. Chem. Soc.* **1910**, *26*, 46.
28 W. H. Perkin, Jr., R. Robinson, *Proc. Chem. Soc.* **1910**, *26*, 131.
29 W. H. Perkin, Jr., R. Robinson, *Proc. Chem. Soc.* **1911**, *27*, 101.
30 W. H. Perkin, Jr., R. Robinson, *J. Chem. Soc.* **1911**, *99*, 775.
31 J. A. Berson, in *Chemical Creativity: Ideas from the work of Woodward, Hückel, Meerwein, and Others.*, Weinheim, **1999**. p. 110–124.
32 R. Robinson, in *Memoirs of a Minor Prophet: 70 Years of Organic Chemistry*, Elsevier, Amsterdam, **1976**. v. 1 p. 25.
33 A. Pictet, F. W. Kay, *Ber.* **1909**, *42*, 1973.
34 A. Pictet, M. Finkelstein, *Ber.* **1909**, *42*, 1979. In fact, the first in vitro synthesis of laudanosine had been reported a year earlier by these authors, who however had not used the Mannich-type aldehyde route but instead had accomplished the ring-closure by a modified Bischler-Napieralski cyclization of the homoveratroylamide of the dimethoxylated phenylethylamine. This led of course to a dihydro (not a tetrahydro) isoquinoline, which then was reduced in a separate step (see Scheme 8.11).
35 R. Robinson, in *Memoirs of a Minor Prophet: 70 Years of Organic Chemistry*, Elsevier, Amsterdam, **1976**. v 1 p. 110.
36 H. L. Holmes, in *The Alkaloids*, (Eds.: R. H. F. Manske, H. L. Holmes) Academic Press, New York, **1950**. I p. 375ff.
37 H. L. Holmes, in *The Alkaloids*, (Eds.: R. H. F. Manske, H. L. Holmes) Academic Press, New York, **1952**. II p. 513ff.
38 H. L. Holmes, R. Robinson, *J. Chem. Soc.* **1939**, 603.
39 J. A. Berson, in *Chemical Creativity: Ideas from the work of Woodward, Hückel, Meerwein, and Others.*, Weinheim, **1999**. p. 116ff. A review.
40 R. Willstätter, in *Aus Meinem Leben*, Verlag Chemie, Weinheim, **1949**. p. 133. "Baeyer hat niemals ängstlich vermieden, was so viele von uns für unwissenschaftlich halten, eine Hypothese auszusprechen, die experimentell ungenügend gestützt und nur für kurze Lebensdauer bestimmt war. Er brauchte die Hypothese in ihrer Wandelbarkeit. Etwas ungemein Wichtiges aber und sogar Engültiges waren ihm die Formeln der Structurchemie, die ja zu einem grossen Teil ihres Inhalts von den Theorien unserer Zeit unabhängig und dafür bestimmt sind, auch bei tiefgreifenden Umwälzung unserer Anschauungen in neue Ausdrucksformen übertragen zu werden."
41 L. H. Briggs, H. T. Openshaw, R. Robinson, *J. Chem. Soc.* **1946**, 903.

42 R. Robinson, *Nature* **1947**, *159*, 263.
43 R. N. Chakravarti, R. Robinson, *Nature* **1947**, *160*, 18.
44 R. B. Woodward, **1946**: Personal communication to Robinson cited in the preceding reference.
45 R. B. Woodward, W. J. Brehm, *J. Am. Chem. Soc.* **1948**, *70*, 2107.
46 R. B. Woodward, W. J. Brehm, A. L. Nelson, *J. Am. Chem. Soc.* **1947**, *69*, 2250.
47 H. Leuchs, *Ber.* **1940**, *73*, 731.
48 H. Leuchs, H. G. Boit, *Ber.* **1940**, *73*, 885.
49 H. Leuchs, E. Tuschen, M. Mengelberg, *Ber.* **1944**, *77*, 408.
50 R. B. Woodward, M. P. Cava, W. D. Ollis, A. Hunger, H. U. Daeniker, K. Schenker, *J. Am. Chem. Soc.* **1954**, *76*, 4749.
51 R. B. Woodward, in *XIVth Congress of Pure and Applied Chemistry, Vol. 2*, Birkhaüser Verlag Basel, Zürich, **1955**, p. 21.
52 R. B. Woodward, M. P. Cava, W. D. Ollis, A. Hunger, H. U. Daeniker, K. Schenker, *Tetrahedron* **1963**, *19*, 247.
53 R. Robinson, in *Progr. Org. Chem.*, (Ed.: J. W. Cook) Butterworth's, London, **1952.** 1 p. 1.
54 G. Barger, C. Scholz, *Helv. Chim. Acta* **1933**, *16*, 1343.
55 G. Hahn, H. Ludewig, *Ber.* **1934**, *67*, 2031.
56 R. Robinson, *The Structural Relations of Natural Products*, Clarendon Press, Oxford, **1955**.
57 C. Schöpf, H. Beyerle, *Ann.* **1934**, *513*, 190.
58 R. B. Woodward, *Nature* **1948**, *162*, 155.
59 R. Robinson, *Nature* **1948**, *162*, 524.
60 (a) A. J. Birch, in *Perspectives in Organic Chemistry*, (Ed.: A. R. Todd) Interscience, New York, **1956.** p. 134–153. (b) R. Goutarel, M. M. Janot, V. Prelog, W. I. Taylor, *Helv. Chim. Acta*, **1950**, *33*, 150.
61 R. B. Turner, R. B. Woodward, in *The Alkaloids*, (Eds.: R. H. F. Manske, H. L. Holmes) Academic Press, New York, **1953.** III p. 54–63.
62 R. B. Woodward, *Angew. Chem.* **1956**, *68*, 13.
63 R. Robinson, in *Festschrift Prof. Dr. Artur Stoll zum siebzigten Geburtstag*, (Ed.: E. Jucker) Birkhäuser Verlag, Basel, **1957.** p. 457.
64 R. B. Woodward, in *Perspectives in Organic Chemistry*, (Ed.: A. R. Todd) Interscience, Ne w York, **1956.** p. 155–183.
65 D. R. Dalton, in *The Alkaloids: The Fundamental Chemistry – A Biogenetic Approach*, Marcel Dekker, New York, **1979.** p. 216–227., and references cited therein.
66 D. R. Dalton, in *The Alkaloids: The Fundamental Chemistry – A Biogenetic Approach*, Marcel Dekker, New York, **1979.** p. 49–58. Also references cited therein.
67 A. R. Battersby, *Proc. Schem. Soc.* **1963**, 189.
68 D. H. R. Barton, *Proc. Chem. Soc.* **1963**, 293.
69 L. Stryer, *Biochemistry*, W.H. Freeman, San Francisco, **1975**. Chapters 20 and 21 and references cited therein.
70 A. R. Battersby, *Pure Appl. Chem.* **1967**, *14*, 117.
71 A. R. Battersby, in *The Alkaloids*, The Chemical Society, London, **1971.** 1 p. 31–47.
72 D. R. Dalton, *The Alkaloids: The Fundamental Chemistry – A Biogenetic Approach*, Marcel Dekker, New York, **1979.** Also references cited therein.
73 G. Massiot, C. Delaude, in *The Alkaloids Chemistry and Pharmacology*, (Ed.: A. Brossi) Academic Press, San Diego, **1988.** 34 p. 211.
74 A. Rahman, A. Basha, *Biosynthesis of Indole Alkaloids*, Clarendon Press, Oxford, **1983**.
75 A. I. Scott, in *MTP International Review of Science*, (Ed.: K. Wiesner) Butterworths University Park Press, London, **1973.** 9 p. 115ff.
76 A. I. Scott, *Science* **1974**, *184*, 760.
77 P. Loew, H. Goeggel, D. Arigoni, *Chem. Comm.* **1966**, 347.
78 E. Wenkert, N. V. Bringi, *J. Am. Chem. Soc.* **1959**, *81*, 1474.
79 E. Wenkert, *J. Am. Chem. Soc.* **1962**, *84*, 98.
80 R. Thomas, *Tetrahedron Lett.* **1961**, 54.
81 K. Mothes, F. Weygand, D. Gröger, H. Grisebach, *Z. Naturforsch.* **1958**, *136*, 41.
82 A. J. Birch, H. Smith, in *CIBA Foundation Symposium on Amino Acids and Peptides with Antimetabolic Activity*, (Eds.: G. E. W. Wolstenholme, C. M. O'Conner) J. and A. Churchill, London, **1958.** p. 247. A. J. Birch, B. J. McLoughlin, H. Smith, Tetrahedron Lett. **1960**, I.
83 N. Nakagura, G. Hoefle, M. H. Zenk, *J. Chem. Soc. Chem. Commun.* **1978**, 896.

84 A. R. Battersby, R. J. Parry, *J. Chem. Soc. D* **1971**, 901.

85 C. Schlatter, E. E. Waldner, H. Schmid, W. Maier, D. Gröger, *Helv. Chim. Acta* **1969**, *52*, 776; D. Gröger, W. Maier, P. Simchen, *Experientia* **1970**, *26*, 820; D. Gröger, W. Maier, in *Int. Symp.* (Ed.: K. Mothes), Akademische-Verlag, Berlin, Halle, **1969**, p. 379; S. I. Heimberger, A. I. Scott, *J. Chem. Soc., Chem. Commun.* **1973**, 217.

86 K. Sheth, E. Ramstad, J. Wolinsky, *Tetrahedron Lett.* **1961**, 394; A. J. Birch, J. Grimshaw, *J. Chem. Soc.*, **1961**, 1407.

87 W. R. Dunstan, F. W. Short, *Pharm. J. Trans.* **1883**, *14*, 1025.

9
Epilogue

"... logic and intuition each have their necessary role. Each is indispensable. Logic, which alone can give certainty, is the instrument of demonstration; intuition is the instrument of invention."
Henri Poincaré, 1913. [1]

9.1
Where Have We Been?

While reading this book, you will have become aware of certain incongruencies between the admonitions of some philosophers and the events of the history of chemistry. For example, we have been advised to reject theories that are internally contradictory, but Kekulé's theory of benzene, although clearly prohibited by this criterion, still served to guide the development of organic chemistry in the 19th century. Kekulé's attempt to salvage the theory with his collision proposal defied Popper's requirement that a rescuing hypothesis must itself be independently testable. Similarly, one could argue that the whole idea of driving science by conjectures and refutations is called into question by the brilliantly successful but ultimately disproven aromatic fission theory of alkaloid biosynthesis. Again, in the supposed disproof of the transannular hydrogen shift theory in bicyclic terpene rearrangements, we found an example of difficulty in establishing at which point in history we should take refutations as received wisdom. Similarly, the discovery of crystalline proteins undoubtedly shrank what might be called the "credibility-space" remaining to the Willstätter small molecule theory of enzymes, but although it was quickly *seen* as a refutation, in fact it was not.

It is as though chemists operate in a universe parallel to that of an influential segment of the philosophy of science community. In fact, this discrepancy is by no means unique to chemistry but occurs in other sciences as well. What are we to make of this? Do the promulgators of scientific methodological orthodoxy really believe the experiences of the natural sciences to be irrelevant, or are there other factors at work?

9.2
The "Logicians' Program"

Of course, some philosophers also have pondered the question. I think a particularly clear view is that of Stephen Toulmin. [2] He believes that

> "... philosophers of science share a widespread feeling that methods of analysis which have served them well for half a century are reaching the limits of their usefulness. During much of the 20th century, they have been concentrating on those aspects of science which lend themselves to analysis using the tools of formal logic ... the resulting debate has been complex and on a high level of abstraction ... from year to year the argument has moved on, and the grounds of debate have shifted. It has all been very exciting. Very exciting, yes: but also oddly inconclusive – and this is where the current malaise begins."

He argues that since the 1920s, the intellectual course on which philosophers set out, namely the *"logicians' program"* for the philosphy of science, has been completed, for the most part. Yet that work has left many acute and important problems in the philosophy of science unanswered. Thus, he writes,

> "Originally it was hoped to bring all serious philosophy of science within the logicians' ambit by sufficiently ingenious extensions. Now, this appears increasingly unrealistic, and nobody is entirely clear where to turn next."

Toulmin makes what I consider to be a useful distinction between two classes of problems in the methodology of science, which for convenience I call here Class I and Class II. Class I problems concern the structure of arguments in which pre-existing concepts are applied to science. Philosophers analyze these arguments using formal logical principles; they set standards for good science in terms of adherence to these principles. The goal is to carry out accurately the processes of inference from an already known set of axioms or other logical constructs, in other words, to avoid the blunders of illogicality in our scientific reasoning, as Toulmin writes:

> "In the actual *doing* of science, logicality represents a kind of intellectual accountancy: concerned with keeping our inferences neat and tidy, avoiding formal blunders, doing our sums aright ..."

Outside of that domain are the Class II problems. These exist at the methodology-practice interface. They have to do with what Toulmin calls the *rationality* or *reasonableness* of science, with the way in which new concepts are developed in the solution of scientific problems:

> "... this is a matter, not of following established procedures rightly (*i.e.*, impeccably) but of recognizing what are the right (*i.e.*, relevant) procedures to employ in a new situation."

This distinction is not clearly articulated by many of the proponents of the "logicians' program." Their recommendations purport to put a *logical* platform under the whole scientific enterprise. This has led, I believe, to the kinds of mismatches of philosophical guidelines and actual events of science we have recounted in this book. In most cases, these non-superposabilities seem to arise from attempts to apply rules derived from or inspired by Class I considerations to Class II problems.

9.3
Which Way Now?

So is the relationship of science and philosophy destined to remain forever fruitless? Must we join the reviewer [3] of a recent book on philosphy of science, who ended his rather skeptical comments with the question "... I am left wondering, do philosphers really have anything useful to tell scientists?"

I think there may be benefits to both groups potentially available from a continued effort to interact, provided we each recognize where philosophy can help and where it cannot. An understanding of logically derived principles is certainly helpful, and perhaps necessary, to the practice of good science. But "good science" cannot be defined, guided, monitored, or judged by logical rules alone. The experiences of science show that the quality of a theory as evaluated by logical standards is not a guarantor of the fruitfulness of the theory.

Are science and its logically derived methodology fated to be disparate (if not actually orthogonal) branches of learning? Or can we find ways to bridge the gap between science and the logicians' program? I believe that any such rapprochement will require more mutual awareness between the two disciplines.

Whether the practitioners of science can be persuaded to make such an effort is questionable. Traditionally, we scientists think in the short term. If we have a theory that works for the present configuration of facts and predicts new and fruitful experiments, we tend to continue to use it, even though at some point it may be "falsified." We have lived through dozens of "refutations" of our favorite ideas, and we do not worry too much about their lifetimes or their ultimate "truth." We say that we seek "understanding," but we judge whether we have achieved it by whether we can use that knowledge for fruitful guidance on experimentation. Our goals are therefore pragmatic.

Philosophical methodologists, on the other hand, seem to judge "understanding" by deeper or at least different criteria. Their question is epistemological: What is the nature of knowledge and can it be deduced by observation of actual events? In Yeats's words, [4] *"How can we know the dancer from the dance?"* Is it too presumptuous to hope that methodologists will someday see the value of integrating their work more fully with the goals and actual course of science itself? Perhaps the present study and others like it to come can mediate that process.

The progress of chemistry, and I believe, of most of science, takes place in a contexture of chaos. Scientific knowledge grows *despite* illogicality of the means by which it has been acquired, *despite* scientists' clumsiness in manipulating the tools of proper inference. On our journey to understanding, we often congratulate ourselves (usually for only a short time) when we have reached some plateau from which we think we can faintly see the ultimate goal. If we wish, we can go back mentally to reconstruct the details of our prior trajectory, but what is of greater interest to us is how to choose the best route to continue the journey. Is there an alternative pathway to the one we had planned, one which is not yet on any map, one which is more direct and which expands the scope of our vision far beyond our expectation? Are there new goals we have not yet discerned that we should be trying to reach? If philosophers can help us to find such trails, scientists will welcome them with open arms as companions on the quest.

9.4 References

1 J. H. Poincaré, *The Value of Science*, in *The Foundations of Science*, G. B. Halsted,The Science Press, Lancaster, PA, **1913** p. 219. Reprinted in 1946.

2 S. Toulmin, *From Logical Systems to Conceptual Populations*, in *PSA 1970 In Memory of Rudolf Carnap*, (Eds.: R. C. Buck, R.S. Cohen), D. Reidel, Dordrecht, The Netherlands, **1971** p. 552ff.

3 L. Wolpert, *Science* **2002**,*295*, 633. A review of *Science, Truth, and Democracy*, by P. Kitcher, Oxford Univrsity Press, New York, 2001.

4 W. B. Yeats, in *W.B. Yeats the Poems: A New Edition (R.J. Finneran, ed.)*, Macmillan, New York, NY, **1983** p. 217.

Index of Authors

References at the end of the chapters are given in the format (chapter), page of citation, [reference number].

a

Acton, N. (7), 128, [19]
Agassi, J. (7), 122, [2]
Arigoni, D. (8), 166 [77]

b

Bacon, F. (2), 17, [17]; (2), 18, [21–24]
Bader, A. (4), 50, [14–15]
Baeyer, A. (6), 112, [78]
Barbieri, J. (6), 114, [81]
Barger, G. (8), 135, [5]; (8), 155, [54, 55]
Bartley, III, W.W. (3) [36]
Barton, D.H.R. (8), 165, [68]
Basha, A. (8), 165, [67]
Battersby, A.R. (8), 165, [67]; (8), 166, [70, 71]
Bellamy, L.J. (7), 123 [4]
Barger, G. (8), 135, [5]; (8), 155, [54]
Bernal, I. (7), 124 [8]
Bernal, J.D. (7), 124, [8]
Berson, J.A. (1), 1, [7]; (3), 29, [8]; (5), 74, [13]; (5), 76, [15]; (5), 20, [20]; (6), 114, [82]; (8), 133, [3]; (6), 114, [82]; (8), 133, [3]
Beyerle, H. (8), 156, [57]
Bhacca, N.S. (7), 123, [5]
Bhushan, N. (1), 2, [9]
Birch, A.J. (8), 139, [17]; (8), 159, [60]; (8), 170, [82]; (8), 172, [86]
Blicke, F.F. (8), 136, [12]
Bondi, H. (3), 28, [5]
Bowman, R.E. (6), 107, [56]
Bredt, J. (5), 73, [11]
Bricmont,J. (1), 2, [10]
Brigel, G. (6), 112, [75]
Briggs, L.H. (8), 154, [41]
Bringi, N.V. (8), 168, [78]
Brush, S.G. (4), 50, [18]
Brüssoff, L. (5), 72, [10]
Buck, R.C. (1), 9, [28]
Butlerov, A. (6), 106, [47–48]
Byron, G.G. (1), 11, [32]
Butterfield, H. (3), 37, [60]
Bytautas, L. (2), 20, [31–32]; (4), 56, [31]

c

Caldin, E.F. (1), 5, [18]
Carnap, R. (2), 15, [9]
Cartwright, N. (6), 116, [86]
Cava, M.P. (8), 161, [50, 52]
Chakravarti, R.N. (8), 150, [43]
Cherbulicz, E. (8), 135, [6]
Cohen, R.S. (1), 9, [28]
Collins, C.J. (6), 107, [55]
Cordell, G.A. (8), 136, [9]
Couturier, F. (6), 106, [50]

d

Daeniker, H.U. (8), 161, [50, 52]
Dalton, D.R. (8), 136, [8]; (8), 163–164, [65, 66]; (8), 166, [72]
Danovich, D. (4), 62, [39]
de Broglie, L. (2), 19, [25]
DeLaude, C. (8), 166, [73]
Dennler, D. (6), 107, [52]
Djerassi C. (6), 113, [79]
Doering, W. von E. (5), 76, [16]
Doyle, A.C. (6), 112, [77]
Duhem, P. (2), 14, [5–6]; (2), 19, [26], (2), 20, [29], (2), 21, [34], (2), 22, [35]; (2), 23, [36]; (3), 38, [62–63]
Dunstan, W.R. (8), 172, [87]

e
Earman, J. (6), 116, [91]
Eastham, J.F. (6), 113, [80]
Ebersbach, L. (6), 98, [25]
Edmonds, D. (3), 3, [3]
Eidinow, J. (3), 3, [3]

f
Fehér, M. (3), 30, [18],
Feigl, H. (3), 30, [42]
Feyerabend, P. (1), 4, [14–15]; (1), 7, [22]; (3), 30, [17]; (3), 30, [40–41]; (3), 36, [51]; (3), 37, [59];(4), 48, [3]; (5),78, [23]; (7), 127, [14]
Feynman, R.P. (8), 133, [1]
Fichter, F. (6), 97, [21]; (6), 99, [28]; (6), 100–101, [29–30]
Fieser, L.F. (6), 107, [51]; (7), 123, [6]
Fieser, M. (6), 107, [51]; (7), 123, [6]
Finkelstein, M. (8), 146, [34]
Fittig, R. (6), 93, [6]; (6), 97, [22]; (6), 97, [24]; (6), 101, [34]
Fodor, J.A. (6), 16, [89]
Forster, F. (6), 114, [81]
Friedel, C. (6), 102, [36]; (6), 103, [38–46]
Fruton, J.S. (3), 37, [57]; (5), 79, [25]; (5), 82, [39]; (5), 85, [44]
Fuller, S. (2), 15, [7–8]

g
Galison, P. (5), 68, [3]
Gams, A. (8), 143, [25]
Gardner, M. (3), 30, 32, [31]; (5), 77, [22]
Geison, G.L. (7), 124, [8]
Gero, A. (4), 21, [21]
Giere, R.N. (6), 116, [90]
Goeggel, H. (8), 166, [77]
Gouvéa, F.Q. (2), 17, [20]
Gower, B. (1), 1, [3–4]; (1), 2, [8]; (1), 6, [20]; (1), 7, [24]; (1), 8, [25]; (1), 9, [27]; (2), 16, [10–11, 14, 16]; (3), 30, [16], (3), 32, [44]; (7), 122, [3]
Grimaux, E. (6), 109, [67–68]
Grimshaw, J. (8), 172, [86]
Grisebach, H. (8), 170, [81]
Gröger, D. (8), 170, [81]; (8), 171, [85]

h
Hafner, K. (4), 53, [23]
Hahn, G. (8), 155, [55]
Hanson, N.R. (3), 39, [65], (3), 40 [69]; (4), 64, [42]
Heimberger, S.I. (8), 171, [85]
Hempel, G.C. (2), 17, [18–19]; (3), 30, [15]; (6), 116, [88]
Henry, T.A. (8), 141, [20]
Herriott, R.M. (5), 80, [34–35]
Hesse, M. (1), 6, [19]
Hiberty, P. (4), 62, [39]
Hjelt, E. (6), 94, [12]
Hoefle, G. (8), 171, [83]
Hoffmann, R. (6), 113, [79]
Hofmann, A.W. (7), 124, [8]
Holliday, R. (3), 30, [30]
Holmes, F.L. (6), 113, [79]
Holmes, H.L. (8), 147, [36–37]; (8), 148, [38]
Hopf, H. (4), 56, [29]
Houben, J. (5), 72, [8–9], (5), 73, [12]
Howard, W.L. (6), 95, [16]
Hückel, W. (4), 50, [19]
Huisgen, R. (5), 80, [33]
Hunger, A. (8), 161, [50, 52]

i
Ihde, A. (2), 20, [28]; (2), 20 [30]; (3), 42, [70]; (4), 50, [20]; (6), 94, [11]; (7), 124, [8,9]

j
Jardine, L. (2), 16, [12]
Jena, A. (6), 109, [65, 70]
Johnson, L.F. (7), 123, [5]

k
Kane, R. (6), 96, [19]
Katz, T.J. (4), 56, [28]; (7), 128, [19];
Kauffman, G.B. (7), 124; [8]
Kay, F.W. (8), 146, [33]
Kay, I.A. (6), 107, [55]
Kekulé, A. (4), 50, [10–12]; (4), 50, [22]; (4), 56, [30]; (6), 91, [1–2]; (6), 91, [3]; (6), 101, [31]; (6), 109, [71]
Kilmister, C.W. (3), 28, [5]
Kitcher, P. (1), 9, [29–30]; (1), 10, [31]
Klein, D.J. (2), 20, [31–32]; (4), 56, [30–31]
Kmet, T.J. (6), 107, [56–7]
Kuhn, T.S. (1), 7, [21, 23]; (3), 30, [12]; (3), 35, [53]; (4), 64, [43]; (5), 68, [2]; (6), 113, [79]
Kunitz, M. (5), 80, [35]; (5), 82, [40]
Kurz, E.M. (3), 30, [26]

l
Labinger, L. (1), 3, [12]
Lachman, A. (6), 109, [66]
Ladenburg, A. (4), 53, [26]
Lakatos, I. (1), 9, [28]; (3), 30, [14]; (3), 30, 36, [34]; (3), 30, [37], (3), 35, [52, 56]; (4), 48, [4–5]; (4), 49, [7]; (4), 60, [37]; (7), 122, [1]
Laurent, A. (6), 108, [62]
Lee, C.C. (5), 76, [17]

Lenin, V.I. (3) [59]
Leuchs, H. (8), 151, [47–49]
Liebig, J. (6), 96, [17–18]; (6), 108, [61]
Limpricht H. (6), 101, [30, 33]; (6), 109, [64, 69]
Linnemann, E. (6), 102, [37]
Lipton, P. (6), 116, [84]
Litten, F. (5), 80, [32]
Loew, P. (8), 166, [77]
Löwig, C. (6), 96, [20]
Ludewig, H. (8), 155, [55]

m

Mach, E. (2), 14, [4]
Maier, W. (8), 171, [85]
Mannich, C. (8), 137, [14–15],
Margenau, H. (3), 35, [54]
Massiot, G. (8), 166, [73]
Maull, N. (3), 37, [58]
Maxwell, J.C. (3), 42, [72]
McBride, J.M. (2), 20, [33]; (4), 56, [32]
McKay, W.B. (6), 107, [52]
McKenzie, A. (6), 107, [52–53]
McLoughlin, B.J. (8), 172, [86]
Medawar, P. (1), 2, [11]
Meerwein, H. (5), 69, [4–5]; (5), 70, [6]; 72, [7]
Mendius, O. (6), 97, [23]
Meyer, V. (6), 114, [81]
Michl, J. (5), 77, [21]
Miller, D. (3), 30, [32]
Mislow, K. (6), 107, [54]
Molinari, E. (4), 53[27]; (4), 60, [35–36]
Montfort, F. (5), 72, [7]
Mortimer, H. (3), 30, [33]
Mothes, K. (8), 170, [81]
Mott, P. (6), 116, [87]
Myers, R.D. (7), 124, [8]

n

Nakagura, N. (8), 171, [83]
Nakamura, K. (6), 107, [58]
Nametkin, S. (5), 72, [10]
Noe, C. (4), 50, [15]
Northrop, J.H. (5), 80, [34–35]; (5), 80, [40–42]

o

Ollis, W.D. (8), 161, [50, 52]
Openshaw, H.T. (8), 151, [41]
Oppenheim A. (6), 111, [72]
Osamura, Y. (6), 107, [58]
Owen, L.N. (5), 76, [14]

p

Papineau, D. (3), 30, [21–22]
Parry, R.J. (8), 171, [84]
Pawlow, D. (6), 106, [49]
Perkin, Jr. W.H. (8), 144, [27–30]
Perry, Jr. R. (5), 76, [19]
Persky, J. (6), 116, [83]
Pfankuch, E. (5), 72, [8–9]; (5), 73, [12]
Pictet, A. (8), 135; [4], (8), 135, [7]; (8), 136, [10]; (8), 143, [26]; (8), 146, [33–34]
Poincaré, J.H. (1), 1, [1–2]; (1), 8, [26]; (2), 14 [1–2]; (2), 20, [27]; (9), 178, [1]
Polanyi, M. (1), 5, [17]
Popper, K.R. (1), 4, [13]; (2), 24, [37]; (3), 26, [1–2]; (3), 27, [4], (3), 29, [9–10]; (3), 30, [38–39]; (3), 32, [43–45], (3), 34, [47–50]; (4), 47, [1–2]; (4), 48, [6]; (4), 49, [8–9]; (4), 64, [41]; (5), 68, [1]; (7), 128, [15]
Preston, J. (3), 30, [22–24]
Prout, W. (3), 42, [71]
Psarros, N. (1), 1, [6]
Psimopoulos, M. (3), 30, [25]

r

Rahman, A. (8), 166, [74]
Rainey, W.T. (6), 107 [55]
Ramstad, E. (8), 172, [86]
Reichert, B. (8), 136, [11]
Reynolds, J. (5), 79, [26]
Roberts, J. (6), 116, [91]
Roberts, J.D. (5), 76, [17–18]
Roberts, R.M. (7), 7, [7]
Robinson R. (8), 140, [18–19]; (8), 142, [23–24]; (8), 144, [27–30]; (8), 146, [35]; (8), 150, [41–43]; (8), 154,[53]; (8), 155, [56]; (8), 158, [59]
Rocke, A.J. (4), 52, [17]; (4), 63, [40]; (5), 88, [46]; (6), 91, [2]; (6), 92, [4]; (6), 94, [10]; (6), 94, [14]; (6), 95, [15]; (6), 99, [27], (6) [33]
Roger, R. (6), 107, [53]
Rohdewald, M. (5), 80, [28]
Roscoe, H.E. (6), 93, [8–9]
Rosenfeld, S. (1), 2, [9]
Royce, J. (2), 14, [3]
Rudofsky, S. (4), 52, [16]

s

Savary, C. (3), 30, [19]
Schaefer, H. (8), 137, [13]
Schenker, K. (8), 161, [50, 52]
Schilpp, P. (3), 38, [62]
Schlatter, C. (8), 171, [85]
Schmid, H. (8), 171, [85]
Scholz, C. (8), 155, [54]

Schöpf, C. (8), 156, [57]
Schorlemmer, C. (6), 93, [8–9]
Schütt, H.-W. (7), 124, [8]
Schwanert, H. (6), 109, [64, 69]
Scott, A.I. (8), 166, [75–76]; (8), 171, [85]
Selman, S. (6), 113, [80]
Shaik, S. (4), 62, [38–39]
Shedlovsky, T. (5), 82, [42]
Sheth, K. (8), 171, [86]
Shoolery, J.N. (7), 123 [5]
Short, F.W. (8), 172, [87]
Shurki, A. (4), 62, [38–39]
Siegel, M. (6), 107, [54]
Simchen, P. (8), 171, [85]
Simmonds, S. (5), 82, [39]; (5), 85, [44]
Simonsen, J. (5), 76, [14]
Slater, J.C. (3), 35, [55]
Small, L. (8), 141, [21]
Smart, J.J.C. (3), 30, [13]
Smith, H. (8), 170, [82]; (8), 172, [86]
Smith, W.B. (6), 107, [55–58]
Soddy, F. (3), 43, [73]
Solomon, J.R. (2), 16, [13]
Sorensen, S.P.L. (5), 82, [43]
Städeler, G. (6), 98, [26]; (6), 111, [73]
Stryer, L. (5), 85, [45]; (8), 166, [69]
Sumner, J.B. (5), 80, [36]
Sundberg, C. (5), 80, [29]
Symons, R. (6), 112, [76]

t

Tanford, C. (5), 79, [26]
Tauber, H. (5), 80, [38]
Theocharis, T. (3), 30, [25]
Thomas, R. (8), 168, [80]
Tollens, B. (8), 137, [13]
Toulmin, S. (1), 5, [16]; (3), 30, [11]; (9), 179, [2]
Trier, G. (8), 138, [16]
Turner, R.B. (8), 160, [61]

v

van Emster, K. (5), 69, [4,5]
van't Hoff, J.H. (4), 56, [33]
Vaughan, W.R. (5), 76, [19]
von Uslar, L. (6), 101, [30]

w

Waldner, E.E. (8), 171, [85]
Waldschmidt-Leitz, E. (5), 79, [27]
Wallau, W.M. (3), 34, [7]
Wallis, E.S. (6), 93, [5]
Watkins, J. (3), 30 [27]
Weidmann, S. (6) [21]
Wenkert, (8), 168, [78–79]
Weygand, F. (8), 170, [81]
Willstätter, R. (5), 78, [24]; (5), 79, [28]; (5), 80, [30]; (8), 141, [22]
Winterstein, E. (8), 138, [16]
Witkop, B. (5), 80, [31]
Wöhler, F. (7), 124, [10]
Wolf, A.P. (5), 76, [16]
Wolinsky, J. (8), 171, [86]
Wolpert, L. (9), 180, [3]
Woodward R.B. (8), 150, [44]; (8), 151, [45–46]; (8), 154 [50–52]; (8), 158, [58]; (8), 160, [61–62]; (8), 161, [64]
Wortmann, R. (5), 70, [6]
Wotiz, J. (4), 50, [13, 17]

y

Yancey, J.A. (5), 76, [18]
Yeats, W.B. (9), 180, [4]

z

Zenk, M.H. (8), 171, [83]
Zincke, T. (6), 111, [74], (6), 112, [76]
Zinin, N. (6), 108, [63]

Index of Names

Names mentioned in the text are listed as (chapter), page number.

a

Agassi, J. (7), 122
Aristotle (2), 18
Aston, F.W. (3), 42

b

Bacon, F. (1), 9, 11; (2), 16ff; (3), 26, 29
Baeyer, A. (4), 58 ; (6), 96, 112–113; (8), 149
Barger, G. (8), 155–156
Berzelius, J.J. (6), 95
Birch, A.J. (8), 139–140, 144, 159, 170
Bischler, A. (8), 146
Blount, B.K. (8), 148
Bohr (1), 5; (3), 35; (4)
Brehm, W.J. (8), 151
Brigel, G. (6), 112
Bringi, N.V. (8), 168
Brüssoff, L. (5), 72
Butlerov, A. (6), 95, 104–107, 105 (Figure 6.3), 113–115
Butterfield, H. (3), 37
Byron, G.G. (1), 11

c

Carnap, R. (1), 9, 28; (3), 32
Caro, H. (6), 112
Cartwright, N. (6), 119
Chakravarti, R.N. (8) 150–151
Chenevix, R. (6), 96
Chou, T.Q. (8), 136
Couper, A.S. (6), 95, 115
Couturier, F. (6), 106
Crookes, W. (3), 43

d

de Broglie, L. (2), 23
Descartes, R. (1), 5; (3), 29
Dewar, M.J.S. (8), 158
Djerassi, C. (6), 121 (in fn. 79)
Doering, W. von E. (8), 162)
Duhem, P. (1), 11; (2), 14, 19ff; (3), 26, 28, 37ff, (8), 133
Dulong, P.L. (6), 95
Dumas, J.P.A. (6), 95
Duns Scotus (3), 34

e

Earman, (6), 119
Ebersbach, L. (6), 98
Einstein, A. (1), 5, 9; (2), 17, 23 ;(3) 28; (5), 77
Elizabeth I (2), 16
Empiricus, S., (3), 32

f

Feigl, H. (3), 30
Feynman, R. (8), 133
Feyerabend, P. (1), 4, 11; (3), 30, 35ff, 39 (in Figure 3.2); (4), 48; (5), 77–78
Fichter, F. (6), 97
Finkelstein, M. (8), 146
Fittig, W.R. (6), 92 (Figure 6.1), 93–94, 96ff, 112; (7), 123
Friedel, C. (6), 100 (Figure 6.2), 102–106
Friedländer, P. (6), 112–113
Fruton, J.S. (5), 82
Futaki, R. (5), 77

g

Gams, A. (8), 142–143
Gardner, M. (3), 32; (5), 77
Gerhardt, C.F. (6), 95
Gero, A. (4), 58, 61
Gibbs, J.W. (5), 82, 84

Gomberg, M. (5), 84 (Figure 5.4)
Goutarel, R. (8), 159
Gower, B. (1), 7ff, 11; (2), 16; (7), 122
Grimaux, E. (6), 109–110

h

Hahn, G. (8), 155–156
Hanson, N.R. (3), 39ff
Hesse, M. (1), 6
Hempel, G.C. (2), 17; (6), 116
Hilbert, D. (2), 17
Hjelt, E. (6), 94
Hobbes, T. (3), 29
Hoffmann, J.H. (6), 97
Hoffmann, R. (5), 77; (6), 121 (in fn. 79)
Hollitscher, W. (3), 37
Holmes, F.L. (6), 121 (in fn. 79)
Holmes, H.L. (8), 147
Holmes, S. (6), 112
Houben, J. (5), 72ff
Hückel, W. (4), 58
Hückel, E. (3), 34; (4), 61; (6), 118
Hume, D. (3), 29

i

Ihde, A. (3), 42; (6), 94
Ipatieff, V. (5), 84 (figure 5.4)

j

James I (2), 16
Janot, M.M. (8), 159
Jena, A. (6), 110–112

k

Kane, R. (6), 96
Kant, E. (3), 29
Kekulé, A. (4), 47ff, 50 (in Figure 4.1); (6), 91ff, 95, 102, 110, 114–118; (8), 134; (9), 178
Kitcher, P. (1), 9
Körner, W. (4), 50 (in Figure 4.1), 58
Kraft, V. (3), 36
Krösche, W. (8), 137
Kuhn, T.S. (1), 28; (3), 34, 37ff; (6), 113; (8), 134
Kunitz, M. (5), 82

l

Ladenburg, A. (4), 50 (in Figure 4.1), 58, 63
Lakatos, I. (1), 11; (3), 30, 34ff; (4), 48ff, 62, 64ff; (6), 117; (7), 122
Lapworth, A. (8), 140
Laurent, A. (6), 95; (7), 123
Le Bel, J.A. (2), 20; (4), 56
Leibniz, G.W. (3), 29
Lenin, V.I. (3), 37
Leuchs, H. (8), 147, 151
Liebig, J. (6), 96, 108–109, 113; (7), 123
Limpricht, H. (6), 97–101, 109–111
Linnemann, E. (6), 102
Locke, J. (3), 29
Lorentz, H. (3), 35
Loschmidt, J. (4), 50
Löwig, C. (6), 96

m

Mach, E. (1), 5; (3), 28
Mannich, C. (8), 136ff
March, A. (3), 37
Margenau, H. (3), 34
Maxwell, J.C. (3), 35, 42
Medawar, P. (1), 2
Meerwein, H, (5), 69ff; (8), 134
Menon, K.N. (8), 147
Menten, M. (5), 85
Michaelis, L. (5), 85
Mill, J.S. (3), 29
Molière (pen name of Poquelin, J.B.) (4), 48
Molinari, E. (4), 60
Montfort, F. (5), 72

n

Nagel, E. (3), 33
Nametkin, S. (5), 76
Napieralski, B. (8), 146
Nernst, W. (3), 34
Newton, I. (4), 47; (6), 117
Northrop, J.H. (5), 80, 81 (Figure 5.2)

o

Oosterhoff, L. (5), 77
Oppenheim, A. (6), 110

p

Pascal, B. (2), 22
Passerini, M. (5), 72
Pasteur, L. (7), 123–124
Pauling, L. (6), 119
Pawlow, D. (6), 106
Perkin, W.H., Jr. (8), 140, 145ff
Pfankuch, E. (5), 72
Pictet, A., (8), 135–136, 142ff
Plato (2), 25
Poincaré, J.H. (1), 1, 5, 8;(3), 28, 29; (6), 119
Polanyi, M. (3), 28
Popper (1), 11; (2), 24; (3), 26 (Figure 3.1), 28ff; 44 (fn. 3) (4), 47ff, 64ff; (5), 68, 77; (7), 122, 126–127; (8), 133, 173
Prelog, V. (8), 147, 154, 159, 173

Prout, W. (3), 42
Pyrrho (2), 23

r
Regnault, H.V. (3), 38
Roberts, J. (6), 119
Robinson, R. (8) 140 (Figure 8.1); (8), 140ff
Rocke, A.J. (5), 88; (6), 94, 101
Rohdewald, M. (5), 80
Royce, J. (2), 14
Russell, B. (3), 28, 29

s
Sarett, L. (8), 163
Schöpf, C. (8), 156
Schwanert, H. (6), 110–11
Silva, R. (6), 103
Simmonds, S. (5), 82
Slater, J.C. (3), 34
Snow, C.P. (1), 2
Soddy, F. (3), 43
Sorensen, S.P.L. (5), 82
Spengler, T. (8), 136, 142
St. Louis (Louis IX, King of France) (6), 99
Städeler, G. (6), 98–99, 110–111
Sumner, J.B. (5), 80ff, 81 (Figure 5.1)
Sundberg, C. (5), 80
Symons, R. (6), 112

t
Tauber, H. (5), 80
Taylor, W.I. (8), 159
Thomas, R. (8), 168ff
Thomson, J.J. (5), 85
Tollens, B. (8), 137
Toulmin, S. (9), 179
Trier, G. (8), 138ff
Turner, R.B. (8), 160

v
van Emster, K. (5), 69
van't Hoff, J.H. (2), 20; (4), 56

w
Wagner, G. (8), 149
Watson, J. (6), 112
Weidmann, S. (6), 96
Wenkert, E. (8), 166–168ff
Weyl, H. (1), 5
Wheland, G.W. (6), 119
Whitehead, A.N. (3), 28
Wieland, H. (8), 147
Williamson, A.W. (5), 88
Willstätter, R. (5), 78ff, 84 (Figure 5.4), (7), 128; (8), 141, 149
Winterstein, E.H. (8), 138ff
Wittgenstein, L. (3), 28; (3), 44 (fn. 3)
Wöhler, F. (6), 97, 100–101; (7), 123–126
Woodward, R.B. (5), 77; (8), 134, 147, 150ff, 152 (Figure 8.2)
Wortmann, R. (5), 70, 72ff
Wurtz, A. (6), 102

y
Yeats, W.B. (9), 180

z
Zincke, T. (6), 111–113
Zinin, N. (6), 108

Index of Subjects

a

accidental discoveries
  examples of, 122ff
  Popper's view of, 126ff
acetone
  elemental composition of, 95
  Fittig's reduction of in attempt to alkylate, 96ff
  structure of, 95ff
alkaloids
  biosynthesis of, 133ff, 166ff
atomic weight controversy
  and experimental determination of molecular structure, 96ff
  and Karlsruhe Conference, 95
  and structural theory, 95

b

Bacon, Francis
  and crucial experiment, 18ff
  and induction, 14ff
  criticism of Aristotle, 18ff
Baeyer, Adolf von
  and benzilic acid rearrangement, 110ff
  and Wagner rearrangement, 149
  criticism of Kekulé's benzene formula, 53, 66 (fn. 27)
  reverence for structural formulas, 149
benzene, *see* Kekulé's theory
benzilic acid rearrangement, 108ff
berberine bridge
  in biosynthesis of berberine, 165
  in biosynthesis of yohimbine, 155
biogenesis of terpenes, iridoids, steroids, amino acids, and alkaloids, 166ff
biogenesis of alkaloids
  experimental tests of Robinson's theories, 163ff
  experimental tests of the Wenkert-Thomas hypothesis, 170ff
  experimental tests of the Woodward fission hypothesis, 170ff, *see also* Woodward fission hypothesis
  pervaded by the Mannich reaction, 136ff
  Pictet's theory of alkaloids from amino acids, 134
  Robinson's theory, *see* Robinson
  Wenkert-Thomas hypothesis, 166ff
  Winterstein and Trier, 139
Butlerov, Alexander Mikhailovitch
  elucidation of the pinacol rearrangement, 105ff
  structure of pinacolinic acid, 105
  structure and synthesis of pinacol, 106

c

case studies, 6
  use in testing methodology, 6–9
ceteris paribus
  and completeness of theories, 119
  as distinct from provisos, 116
  disregard of, 119
confirmation, *see* corroboration
conjecture:
  source of hypothesis or theory, 27ff
corroboration
  as motivation for synthesis, 106, 128, 143, 156
  Popper's criticism of as test of theory, 31ff
crucial experiment
  Bacon and, 18ff
  crystallization of enzymes, 80ff
  Meerwein's transannular shift, 69

d

deductive method
  and Popper's method, 27ff
  as basis of falsificationism, 11

discovery
  importance of proper assignment of credit for, 114ff
  Kuhn's definition of, 113
  of benzilic acid rearrangement, 108ff
  of pinacol rearrangement, 83ff
Duhem, Pierre
  and chemistry, 20ff
  and conviction, 22
  and natural classification, 19ff
  and reasons of the heart, 22
  criteria for a physical theory, 19ff
  criterion of truth of theory, 21
  criticism of quantum nature of light, 23

e

empiricism, 8–9
enzymes
  as proteins, 78
  crystallization of, 80
  limit of reaction rates, 86ff
  Michaelis-Menten kinetics in reactions of, 85
  phase rule test of purity, 82ff
  purity of, 80ff
  Willstätter's small molecule theory, 79ff
experiment
  and incommensurability, 38ff
  entwined with theory, 38ff
  Francis Bacon and, 16ff

f

falsifiability in principle, 64ff
falsification, see also refutation
  step in test of theory, 27ff
Feyerabend, Paul
  and ruthless opportunism, 37
  anything goes, 37
  criticism of Popper's methodology, 30ff, 35, 36ff
  experiment entwined with theory, 39
  personal relationship with Popper, 36
  photo figure, 36
Fittig, Wilhelm Rudolf
  adherence to conventional atomic weights, 98
  at Göttingen, 97
  disagreement with Städeler, 98
  observes rearrangement of pinacol to pinacolone, 99ff
  relationship with Limpricht; 97ff
  student of Limpricht, 97ff
Friedel, Charles
  attempts to elucidate the pinacol rearrangement, 102ff
  deduces structure of pinacol, 103
  incorrect structures for pinacolone and pinacolinic acid, 103ff
  resists molecular rearrangement, 104ff

g

Gardner, Martin
  criticism of falsificationism, 12ff, 77
gestalt diagrams
  photofigures, 40ff

h

Hanson, N.R.
  and perception, 39ff
  criticism of falsificationism, 64

i

inductive method
  Francis Bacon and, 14ff
internal contradiction
  fatal flaw of theory, 34ff
  logical unacceptability of, 34, 39ff, 49ff
iridoids
  as intermediates in biosynthesis, 168ff
  bioynthesis of, 166ff

k

Kekulé, August (von Stradonitz)
  and benzilic acid rearrangement, 110ff
  and students at Ghent photo figure, 50
  influence on Fittig, 99
  rules of chemical reaction, 91ff
  see theory of benzene, 50ff
  textbook, 91, 99
Kekulé's theory of benzene
  atomic collision hypothesis, 61ff
  internally inconsistent, 56
  Ladenburg's prismane; 53, 56ff
  rescue by an ad hoc hypothesis, 62
  structure of benzene, 50ff
  success of despite ad hoc rescue, 62ff
  tests by number of substituted derivatives, 52ff
Körner, Wilhelm
  and criticism of Kekulé's benzene formula, 53
  photo figure, 50
Kuhn, Thomas
  and incommensurability, 39
  and perception, 40

disagreement with Popper, 43, 64
on abandonment of refuted theories, 68

l

Ladenburg, Albert
criticism of Kekulé's benzene theory, 53ff
photo figure, 50
prismane formula, 53
Lakatos, Imre
and contradictory grafts to save theories, 35
and experimental tests of theory, 122
and positive heuristic, 64ff
and Prout's theory, 42
and provisos, 117
attempt to salvage falsificationism, 49
criticisms of Popper's falsificationism, 30, 42, 48, 122
theory entangled with experiment, 42ff
laws of nature
and Duhem, 19ff
and Einstein, 28
Liebig, Justus von
and conflict with Laurent, 108
criticism of Laurent unjustified, 109
transformation of benzil to benzilic acid, 108
loganin
in biosynthesis of alkaloids, 172ff
logicians' program, 179ff

m

Mannich reaction
in biosynthesis of alkaloids, 136ff, 155ff, 163ff
strange variant in biosynthesis of berberine, 165
mass defect in cosmic elemental synthesis, 42
mathematics
and reality, 17
as distinct from empirical science, 17
mathematical content of theory, 20
molecular skeletal rearrangement, *see* rearrangement

n

natural classification
and Pierre Duhem, 21ff
recognition of, 22
naturalism, 9
Northrop, John H.
crystallization of pepsin, 82
phase rule method to determine protein purity, 82ff
photo figure, 81

p

Perkin, W.H., Jr
attitude toward reaction mechanisms, 145
mentor of Robinson, 143
synthesis of narcotine, 144
philosophy of science
ambivalence toward in Byron's Manfred, 12
and methodological theory, 7–10
as guideline for scientific practice, 7
pinacol
elemental composition of, 97ff
formation of, 83ff
rearrangement of, 83ff
structure of, 83ff
Pictet, Amé
synthesis of tetrahydroisoquinolines (Pictet-Spengler reaction), 136
theory of alkaloid biogenesis, 134ff
Popper, Sir Karl Raimund
and Wittgenstein, 44 (fn. 3)
attitude toward historical tests of falsificationism, 30ff
criticisms by Feyerabend and Lakatos, 30ff
influence on working scientists, 28
on accidental discoveries, 126ff
on conjectures and refutations, 26ff
on corroboration, 31
on falsificationism, 27ff
on induction, 27ff
photo figure, 26ff
prohibits internal contradictory theories, 34ff, 49ff
rules of scientific method, 29ff
prediction, *see* theory
provisos, *see* ceteris paribus
Prout's theory, 42ff

r

reaction mechanism
camphene rearrangement of the second kind (Nametkin), 72
fenchol to isofenchol rearrangement, 72
in early 20th century, 142, 145
inversion of camphor, 76
Meerwein's transannular (non-vicinal) shift, 69ff, 76
racemization of camphene, 70
Wagner rearrangement, 70

reality
  Poincaré's formulation as thought, 8
rearrangement
  discoveries of 93ff
  Kekulé's rules of chemical reaction, 91
  of benzil, 108ff
  of pinacol, 93
refutation
  by internal contradiction, 47
  illusory or incomplete, 76, 78ff
  not resolved by ad hoc hypothesis, 47ff
  science without, 122
Robinson, Sir Robert
  contributions to the structure of strychnine, 147ff, 150, 154
  dispute with Pictet, 143ff
  photo figure, 140
  professed lack of interest in history of ideas, 144
  relationship with Woodward, 154ff
  spotty citations by, 142
  student of Perkin, 143
  synthesis of tropinone, 140
  theory of alkaloid biogenesis, 140ff

s

science
  a game, 1
  an anarchic enterprise, 37
  ancient, 18
  and truths of nature, 18
  does not rest on bedrock, 24
  epistemology of, 5
  history of, 1
  methodology of, 2
  philosophy of, 1, 5, *see also* philosophy of science
scientific method
  comparison of with practice, 6
  does it exist? 4–5
  guidelines to, 4
scientific progress
  a mounting tide, 22
scientific research
  accidental discoveries in, 123
  correlation in, 122
  motivations for, 122
  synthesis, a corroborative enterprise, 128ff
scientific theory
  demarcation from metaphysics, 27
secologanin
  in biosynthesis of alkaloids, 172ff
steroids
  biosynthesis of, 166ff
Sumner, James B.
  crystallization of urease, 80
  photo figure, 81
synthesis of organic molecules
  as a series of verifications, 130
  as test of theory, 128
  motivations for, 128ff

t

terpenoids
  biosynthesis of, 166ff
theory, *see also* scientific theory
  comparison with experiment, 21
  origin of, 15, 27
  prediction from, 14ff, 55, 65, 75ff, 130
  purpose of, 14
  tests of, 27ff
theory of molecular types
  and mathematical requirement, 20
theory of molecular structure
  and mathematical requirement, 20
tryptamine
  in biosynthesis of indole alkaloids, 155ff
tyrosine
  in biosynthesis of benzylisoquinoline alkaloids, 167

v

verification, *see* corroboration

w

Wenkert-Thomas hypothesis
  in synthesis of alkaloids, 168ff
Willstätter, Richard
  first synthesis of tropinone, 141
  photo figure, 84
  small molecule theory of enzyme structure, 79ff
Woodward, Robert Burns
  and strychnine structure, 151ff
  photo figure, 152
  relationship with Robinson, 154ff
  ring fission hypothesis of biogenesis, 156, 159ff
  total synthesis of strychnine, 161ff
Woodward fission
  application to other alkaloids, 157ff
  immediate acceptance by Robinson, 157
  in biogenesis of strychnine, 156
  in synthesis of strychnine, 161ff

z

Zincke, Theodor
  almost rationalizes the benzilic acid rearrangement, 111ff
  and benzilic acid, 110ff